Simulating the Earth

Simulating the Earth
Experimental Geochemistry

J.R. HOLLOWAY
Arizona State University

B.J. WOOD
Northwestern University

Boston
UNWIN HYMAN
London   Sydney   Wellington

Unwin Hyman Inc.
8 Winchester Place, Winchester, Mass. 01890, USA

Published by the Academic Division of
Unwin Hyman Ltd
15/17 Broadwick Street, London W1V 1FP, UK

Allen & Unwin (Australia) Ltd,
8 Napier Street, North Sydney, NSW 2060, Australia

Allen & Unwin (New Zealand) Ltd in association with the
Port Nicholson Press Ltd,
60 Cambridge Terrace, Wellington, New Zealand

First published in 1988

Library of Congress Cataloging in Publication Data

Holloway, J. R. (John R.)
  Simulating the Earth: experimental geochemistry
1. Geochemistry
I. Title   II. Wood, Bernard
551.9
ISBN 978-0-04-445255-3   ISBN 978-94-011-8028-3 (eBook)
DOI 10.1007/978-94-011-8028-3

British Library Cataloguing in Publication Data

Holloway, John R.
  Simulating the earth: experimental geochemistry/J. R. Holloway &
Bernard Wood.
  p. cm.
Bibliography: p.
Includes index.
ISBN 978-0-04-445255-3
1. Geochemistry.   I. Wood, B. J.   II. Title.
QE515.H72 1988   88-5403
551.9—dc19   CIP

Typeset in 10 on 12 point Times by MCS Salisbury

Contents

Preface

This is a book about the why and how of doing experiments on rocks, minerals, magmas, and fluids. It could have as logically been subtitled "Experimental petrology" as "Experimental geochemistry," but we chose geochemistry to emphasize the broad and overlapping nature of current experimental work.

We have tried to aim the book at a general readership which we hope will include advanced undergraduate students, graduate students, and anyone else interested in learning something about experimental petrology. Although we hope there will be something of interest for the practicing experimentalist, our aim is at the non-experimentalist interested in learning why experiments are useful, what kind of experiments can be done, and what some of the major problems and limitations are and how they can best be avoided. The result of a journey through this book should be an ability to evaluate published experimental work critically and a knowledge of the kinds of problems an experimentalist might be able to help solve. Some details of experimental technique are included in the Appendix for those readers who want to "get their hands dirty." Indeed, one of our main incentives for writing this book was to try to encourage more petrologists and geochemists to become experimentalists.

In our pedagogical approach we have chosen to discuss a small number of case histories as illustrations of principles and techniques. We have tried to select studies we regard as well executed. Fortunately, there are far more good experimental studies published than there was room for in this small volume, and so exclusion of a paper should in no way be viewed as a criticism.

Although we have discussed most aspects of experimental geochemistry, the limitations of space and our own collective knowledge have forced the exclusion of several important areas. These are mainly in the very low pressure and temperature regime and in the very high pressure (>50 kbar) domain. However, our book is not intended to be a compendium of experimental techniques and data. It is rather a statement of our philosophy of experimental geochemistry, with illustrations of how we consider experiments should be done.

We have made no attempt to teach principles of thermodynamics and phase equilibria in this book, and we assume the reader has some background in those subjects.

We thank V. J. Wall for his assistance during planning of the book.

1 The scope

1.1 The questions

If you have just picked up this book for the first time, you may be asking yourself such questions as: What is an experiment? Do experiments tell us anything useful about the Earth? Isn't experimental geochemistry a rather dull cut-and-dried subject?

1.1a What is an experiment?

An experiment involves the laboratory simulation of physical and chemical conditions within the Earth, or at its surface, with the aim of measuring chemical or physical properties of rocks and minerals under these conditions. A wide temperature–pressure range, 0–2,000°C and 1 atm to 50 kbar (equal to 150 km in depth) can be generated experimentally with high precision and accuracy. Higher pressures, greater than 1 Mbar, can be produced with much lower precision (Ch. 3).

The measurements made by experimenters vary widely in scope and difficulty. A typical "run" would involve holding a sample in a sealed capsule at the desired pressure (P) and temperature (T) for a few hours or, at most, for a few days. The sample would then be quenched to 1 atm and room temperature as rapidly as possible and the products examined to determine the stable phase assemblage and the compositions of the coexisting minerals.

Physical properties can also be measured at high P and T. The viscosities and densities of silicate melts are determined by observing the settling rates of platinum spheres (Ch. 9). Density and crystal structure at high pressure and temperature may now be obtained by combining X-ray diffraction techniques with high-pressure anvil devices. Seismic velocities are measured *in situ* at high P and T by ultrasonic interferometry (Ch. 9).

The one thing that cannot be reproduced exactly is time. In the Earth, many processes take millions of years to complete, and even the most patient of us cannot wait that long. What experimenters usually attempt to do, in those cases where time is important, is to determine the equilibrium compositions of minerals, melts, and fluids under the $P-T$ conditions of the experiments. Since equilibrium is the final, static state approached by a chemical system, it is generally assumed that, given the long times involved, geological processes approach the equilibrium state. This assumption has been supported by numerous studies since that of Goldschmidt (1911), who

1

showed that contact metamorphic mineral assemblages in the Oslo area obeyed the phase rule. In experimental systems a number of methods are used to speed up the approach to equilibrium. Starting materials are generally prepared to be as fine-grained as possible in order to maximize the reactive surface area. The reaction may be catalyzed by adding a good transporting agent such as a fluid. Finally, reactions in experimental systems are often studied at higher temperatures than those at which they occur in the Earth. Increasing temperature raises reaction rates and facilitates the attainment of equilibrium. Extrapolation of data to "real" natural conditions is then made with the aid of thermodynamic relationships such as the Clausus–Clapeyron equation.

1.1b   Are experiments useful?

Obviously, the authors consider this a rhetorical question, but it serves to make a point. Experiments are useful only when taken in the context of a large number of other geological measurements: field, mineralogical, seismological, etc. These other measurements generally provide the information that the experimental geochemist is trying to understand. The reader should not be put off by the fact that many experiments are not performed on rocks. Many of the questions are so complex that their answers have to be worked out in simplified chemical systems before the rocks can be experimented upon. Let us consider some of the problems posed by geological data. If rocks exposed at the Earth's surface were to be sampled, they would be found to contain considerable chemical and physical information about the processes involved in their genesis. Volcanic rocks commonly contain phenocrysts residing in a matrix of glass and of microphenocrysts formed during quenching at the Earth's surface. The compositions of these minerals may, in principle, be used to determine the conditions of depth, temperature, and fluid composition under which they formed. The sizes and forms of the phenocrysts provide further information on the rates of crystallization and of cooling of the magma body from which the rock was derived. In metamorphic rocks, garnet porphyroblasts commonly show complex patterns of chemical zoning which are considered to be related to the univariant and divariant reactions involved in garnet production. The zoning, therefore, contains information bearing on the P–T histories of the rocks. The main aim of experimental geochemists is the design and performance of experiments which enable such chemical and physical data to be interpreted quantitatively. There is no way of doing this without experimental results.

A wide variety of crustal and upper-mantle rocks have become exposed at the Earth's surface by a combination of tectonic and volcanic processes. Xenoliths entrained in alkali-basalt and kimberlite magmas provide samples from depths as great as 150 km. The only direct evidence of the min-

2

eralogical constitution of the remaining 6,100 km of Earth radius, however, is in the way it propagates seismic waves. Despite the indirectness of the methods, geophysical data provide important constraints on the interior constituents of the Earth and the other terrestrial planets. Their volumes, masses and moments of inertia are all known. In the case of the Earth, the velocities of propagation of seismic waves have been accurately measured all the way down to the center. Apart from distinguishing crust from mantle from core, the observed increase in velocity with depth in the mantle is known to correlate with an increasing density with depth. This result leads to the question, "Is the increase in density with depth due to phase transitions in a chemically homogeneous mantle, or is that part of the Earth chemically stratified?" The question can be answered only by accurate experimental data on high-pressure phase relationships and on equations of state for the high-pressure minerals which can occur in the deep mantle. Current data are not really good enough for the task, but in view of recent technical advances (Chs. 3 & 9) the collection of appropriate high quality results may be expected in the next few years.

The brief outline given above is intended to demonstrate that experiments are useful and that there are some questions that can be answered only by experiments. The use of experiments to answer questions posed by other geological methods, and the need for integration of experimental, field, petrological, geophysical, etc., data have also been emphasized. Let us now turn to the last question.

1.1c Is experimental geochemistry cut and dried?

Perhaps surprisingly, experimental geochemistry is not cut and dried. There are about as many different approaches as there are geochemists. This statement is best elaborated by discussing some of the different philosophies and emphasizing the one we will concentrate on in this book.

1.2 Experimental philosophy

The geological approach to experimental geochemisty may be summarized as follows: A petrologist has collected a suite of rocks which may, for example, exhibit an interesting subsolidus reaction, or they may be volcanic glasses containing a number of phenocrysts. In order to determine the conditions under which the phenocrysts formed, or the reaction occurred, a couple of typical bulk compositions are selected and experiments are made over a wide $P-T$ range. When the assemblages are duplicated, the experimenter has a good fix on the conditions of crystallization.

This is a perfectly reasonable approach, and in many cases it provides the desired result. It is often unsatisfactory, however, and it can be criticized on

3

a number of grounds. First, in the case of crustal and uppermost mantle solid–solid (fluid absent) reactions, reaction rates are so low that equilibrium cannot be attained under conditions appropriate to the Earth. Data have to be collected at high temperatures and some extrapolation made. There is no obvious or correct method of extrapolation which does not rely entirely on thermodynamic models of the mineral, melt and fluid phases. Second, it is extremely difficult to demonstrate equilibrium in experiments performed on 10-component natural systems. All of the phases are multicomponent and one has, in principle, to prove equilibrium partitioning with respect to all components in order to demonstrate equilibrium. That is why so many studies are performed on analog systems of only three or four components in which equilibrium can readily be proven. Finally, even if a set of experiments on a given rock all reach equilibrium, the results apply only to that rock; they are not applicable to any other rock composition and can only be extrapolated to other bulk compositions by using thermodynamic models for the different phases.

Since experiments are extremely time consuming, it seems unsatisfactory to perform specific studies such as those outlined above if the same effort can be expended to obtain data of general applicability. More general experiments would be those aimed at the construction and testing of thermodynamic models for the mineral, melt and fluid phases of geological interest. Models, not just thermodynamic ones, are very important in all areas of the Earth sciences. Without them, one cannot hope to understand phase equilibria in natural systems because many parts of $P-T$ composition space are difficult for the experimenter to work in. A model is also needed to take care of the time element in geology. Generally this is modeled by treating the rocks as if they had approached chemical equilibrium at some time in their history. Seismologists model the Earth as if it were elastic whereas fluid dynamicists treat it as a viscous fluid. These models both work adequately under some conditions and serve to illustrate the point that a model can be simple and not absolutely correct as long as it helps geologists understand processes within the Earth.

An approach to phase-equilibrium experiments which embodies most of these ideas may be outlined in the following way. The experimenter collects together all of the thermodynamic data, enthalpy H, entropy S, heat capacity C_p, volume V, thermal expansibility α and compressibility β for pure 1-component phases of geological interest (forsterite, albite, anorthite, $NaAlSi_3O_8$ liquid, SiO_2 liquid, H_2O, CO_2 and so on). Phase relationships involving these pure phases in model 2-, 3-, or 4-component systems are then determined. This step serves two purposes. It gives direct semi-quantitative measurement of geologically relevant phase equilibria, and it enables refinement of the thermodynamic database, particularly the relatively poorly constrained enthalpies. This is followed by experiments to determine mixing properties of multicomponent phases such as H_2O-CO_2

fluid, forsterite–fayalite olivine, $CaMgSi_2O_6$–$NaAlSi_3O_8$–$CaAl_2Si_2O_8$ liquids, etc. Since there are a number of simple theoretical treatments which can be used to extrapolate and interpolate such results, they provide the bases of thermodynamic models for the mixed phases which are of most geological interest. Finally, experiments on rocks are used to test the applicability of the models and their validity for extrapolation in $P-T-X$ (composition) space. Complex system data might also be used for small refinements of the models. By building up from simple systems in this manner, one obtains data and tests models which, in principle, permit calculation of phase relations under any desired conditions of pressure, temperature and composition. Provided the models work satisfactorily under arbitrarily imposed conditions, one can confidently predict phase relations for uninvestigated compositions and for parts of $P-T$ space which are not readily accessible to the experimenter.

This approach is not a pie-in-the-sky dream. There are already sufficient well determined experimental data in the literature to enable construction of good models for, at the very least, haplogranitic melts (Burnham 1981) and for most solid phases in natural mafic and ultramafic compositions (Wood & Holloway 1984, Wood 1987, Ch. 4). Much of the rest of this book is devoted to a description of how good experiments are performed and of how they are used to build and test models.

1.3 Conclusions

We began this chapter with three questions, and we wish to end it with three more, together with a reiteration of the importance of geological constraints on any experimental study. Before embarking on a difficult and time-consuming study, the experimenter should ask:

(a) "What problem am I trying to solve?"
(b) "Is it solvable with the techniques at my disposal?"

If the answer to the second is, "I don't know," then the third question is:

(c) "How can I test whether the problem can be solved?"

If (a), (b), and (c) are satisfactorily answered, experiments can begin. Before going too far with the data however, it is necessary to be sure that the results obey the following criteria:

(a) They must measure the things that they set out to measure, e.g., if equilibrium relationships are sought, then equilibrium has to be demonstrated.
(b) They should be, as far as possible, consistent with repeatedly observed geological relationships, e.g., it would be reasonable to find that

hornblende—plagioclase assemblages are stable under the same physical conditions as quartz—biotite—K feldspar—sillimanite. The latter would not be consistent with assemblages of ablite—chlorite—epidote—actinolite, however.

The experimenter who is able to answer (a), (b), and (c) satisfactorily, and to fit in the appropriate geological constraints, is some part of the way towards performing a useful study.

2 The good experiment

2.1 Introduction

Doing almost any experiment requires much effort and expense. There is no point in spending a lot of time and money unless the result really helps solve some geological problem. This means setting up the experiment to minimize uncertainties in the technique itself. But, because there is usually a trade-off between time and precision, it is also important to know the level of precision and accuracy required of the experiment before beginning. In this chapter we discuss what we think are important things to do, and to avoid, in the process of planning and executing experiments.

2.2 The conditions

The most important first step is to decide exactly why one is doing an experiment. There are certainly a few "qualitative" experiments left to be done in which the only purpose is to determine approximately what happens to some starting material subjected to some given P and T. Most such experiments have been done, however, and their result has always been to provoke more questions than they answer. The most useful experiment is likely to be one chosen to answer a very specific question. For instance, how much does replacement of 20 mol% of the hydroxyl sites in phlogopite with fluorine increase its melting temperature? We feel strongly that the best way to set up experiments to answer such questions as these is to begin by building a thermodynamic model and then to design the experiments to test, extend, or replace that model. The model itself should be chosen using as much geological and geochemical background as possible. There would, for example, be no point in choosing to test the transport properties of bromium selenide because there is no likelihood that this is an important phase in ature.

2.2.1 Sensitivity analysis

The question here is, "How much control?" How closely do run conditions have to be controlled and measured in order to answer the questions? These run conditions are pressure (P) and temperature (T), and activities of oxygen, water, etc. The answer depends on details of the reaction such as the relative entropy and volume changes. This is where a thermodynamic

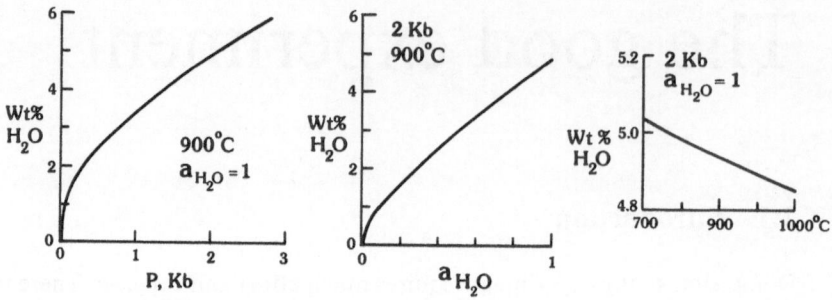

Figure 2.1 The effect of P, T and H_2O activity on H_2O solubility in a rhyolite melt. Calculated using the model of Burnham (1979a).

model is really important because it can be used to estimate the sensitivity of the reaction to the different run parameters. For example Burnham's (1979a) model for the solubility of H_2O in magmas predicts that it depends greatly on P and H_2O activity, but very little on T (Fig. 2.1). From Figure 2.1 we can see that the average change in dissolved H_2O with P is about 2 wt% per kbar, so the P error typical of medium P vessels (± 10 bar) would yield a negligible error in solubility, while the P error typical of higher P machines (± 200 bar) would result in a ± 0.4 wt% error in solubility. The effect of H_2O activity on solubility is about 0.5 wt% for an 0.1 change in H_2O activity. H_2O activities can usually be controlled with a precision of ± 0.01, an uncertainty which results in a negligible error. Finally one can see that the change in solubility with T is only 0.4 wt% per 1,000 degrees, so typical T errors of ± 10 degrees would have no effect on the precision of solubility measurements. Thus, in experimentally determining H_2O solubilities in melts, accurate and precise control of P and H_2O activity is important, but large uncertainties in T can be tolerated. On the other hand, reactions involving oxidation or reduction are far more sensitive to oxygen activity and temperature than they are to pressure.

Once the problem has been defined and the necessary precision and accuracy of the measurements established, the experimentalist is ready to think about some of the mechanics of good experiments. In this chapter we will discuss making starting materials, P and T uncertainties, and other techniques and problems common to most experiments. In the next chapter we will cover the experimental equipment itself.

2.3 What goes in: starting materials

Doing hundreds of runs under carefully controlled $P-T$ conditions will lead to naught if they are done on a poorly prepared starting material. Choosing the right starting material is as important to the experimentalist as it was to Michelangelo when he chose a block of Carrara marble for a sculpture. It is said that he often spent weeks or years deciding on a piece for a particular statue. Many experimentalists wish they had been as careful in their preparations. The reason is that the physical state of the starting material often controls the identity of run products. A wide choice is available: in experiments on rock or mineral compositions the starting material can be a ground powder of a natural material, a glass made by melting that material, a synthetic gel made by precipitation from solution or by freeze-drying, or a synthetic crystalline assemblage.

Fyfe (1960) discussed the effects of starting materials on experimental results and showed that high-entropy starting material usually yields a series of progressively more stable phases, regardless of the relationship of the actual run conditions to the true equilibrium phase boundaries. For example, hydrothermal treatment of amorphous silica yields metastable cristobalite before quartz at any T in the quartz stability field. Once formed, the cristobalite may take months or years to convert to quartz even hundreds of degrees outside its stability field. In addition to the type of material (glass, gel, etc.) its treatment before use can also have a large effect. Newton (1966) showed that the amount of grinding changes the apparent stabilities of sillimanite and kyanite. In a classic paper, Schairer (1959) showed that the heat treatment of a glass can determine whether or not it will crystallize at all.

The important point is that the nature of the experiment dictates the type of the starting material to be used. For determining equilibrium phase assemblages under subsolidus conditions the best material is usually crystalline and contains both the high and low T assemblages. In order to synthesize a single-phase assemblage, the best starting material may be a gel, while to determine garnet stability in mafic systems, a rock powder seeded with garnet crystals works well. More detailed examples of the uses of starting materials will be given in the case histories in later chapters.

2.4 Measurement imperfections

Experiments performed at high T and P are plagued with uncertainties of many kinds. This is what makes them challenging. The objective is to outsmart the system and keep the uncertainties small enough to learn something. In this section we treat the problems of the experiment itself and in the following section the analysis of the sample after the end of the run.

The most obvious uncertainty is in the measurement P and T. In this context it is important to distinguish between precision and accuracy. Precision refers to how closely a measurement can be reproduced, while accuracy refers to how closely the measurement approaches "the truth." An

9

example of the difference may be found in the piston-cylinder high-pressure system which is much used for metamorphic and igneous experiments. Careful use of these systems allows the pressure to be reproduced with high precision (about ±0.2 kbar out of 20 kbar). But in many configurations there are systematic effects causing the measured pressure to be as much as 3 kbar above the actual ("true") pressure. Under most circumstances a high degree of precision is required but high accuracy may not be as important. Careful calibration can be used to correct for systematic errors if measurement precision is high. In the case of T, the factors most affecting precision are gradients in T over the distance between the sample and thermocouple, and drift in the output of the thermocouple with time. The gradients are inherent in the design of the individual equipment (see Ch. 3). Thermocouple drift is generally caused by contamination and can be a terrible problem, but careful choice of thermocouple material and insulators can usually reduce or eliminate the problem. For instance Pt/PtRh thermocouples work well in oxidizing environments in atmospheric pressure furnaces, but are easily contaminated in piston-cylinder furnace assemblies. Conversely, W/WRe thermocouples burn in air at Ts or *temperatures* above 500 °C but work very well in the reducing environment of piston-cylinder furnaces. (Further information on thermocouples is given in the Appendix).

Uncertainties in P and T are caused by the experimental device. An equally serious experimental limitation is imposed by the rate of reaction in the specific system being studied. The experimentalist is often trying to reproduce phenomena that develop over geologic time scales, but is required to produce a result on a much shorter laboratory time scale. For the result to be remotely useful the kinetic limitations have to be carefully considered and overcome. Most of this book is devoted to equilibrium experiments; the expectation of these is that they will have reached equilibrium during the duration of the run, or at least that they will show the direction toward equilibrium. In most that are done, heterogeneous equilibria involving two or more phases are measured. The major factors controlling the reaction rate in these experiments are diffusion, surface reactions, and transport properties of the least viscous phase (usually melt or fluid). Reactions between pure phases (pure quartz for example) are generally controlled by the rate of surface reaction, whereas those involving solid solutions (e.g., olivine of 90% forsterite, 10% fayalite; plagioclase of 30% albite, 70% anorthite) are limited by both surface reaction and the diffusion of the mixed components through the reacting crystals. In either case the rate of reaction is maximized by using starting materials which are as fine grained as possible, because this produces a large surface area for surface reaction and a small distance for interdiffusion.

A completely different approach to determining equilibrium involves the measurement of diffusion, which is a transport property. A well thought out diffusion experiment can often be used to obtain equilibrium data as well as the transport property. For example, Karsten *et al.* (1982) were able to measure H_2O solubilities in rhyolite melts from the "edge values" of the H_2O diffusion profiles used to measure the diffusion coefficient of H_2O.

There are methods of testing the approach to equilibrium. These tests are a necessary and integral part of all experiments aimed at determining equilibrium, and results published without such tests should be treated with caution. The only really certain proof of equilibrium is a reversal experiment in which it is shown that the direction of a reaction can be reversed by changing the P, T, or composition of a reacting system. The melting of albite in the presence of pure H_2O is an example.

Consider the results shown in Table 2.1 taken from Bohlen & Boettcher (1982) for the melting of albite at 10 kbar (± 0.1 kbar) in the presence of a fluid composed of 50% H_2O and 50% CO_2. Starting materials consisted of crystalline albite and a mixture of $Ag_2C_2O_4$ and H_2O.

Runs were done in the order listed. Runs 382 and 390 suggest that albite melting occurs between 790 and 800°C, but do not prove that albite will not melt at 790°C if given more time. To prove that crystalline albite is stable at 790°C a reversal experiment must be done. Bohlen & Boettcher did this in run 455, which was first held at 820°C for 8 hours (conditions which they knew would yield liquid from the results of runs 381 and 390) and then T was lowered to 780°C and held there for 24 hours. The products showed that the liquid crystallized completely. This proves that equilibrium melting occurs between 780 and 800°C.

Let us consider an alternative result to run 455 for illustrative purposes. Suppose the result had instead been the one shown for run 4455, that is, the liquid did not crystallize completely at 780°C. At this point one would have to do another run at lower T, for instance 4456. If its products showed no liquid then one would have a bracket between 760 and 800°C. A narrower

Table 2.1 Results of experiments in the albite–H_2O–CO_2 system.

Run no.	Duration (hours)	T (°C)	Products
382	8	790	albite + fluid
381	8	810	albite + liquid + fluid
390	8.5	800	albite + liquid + fluid
455†	8	820	albite + liquid + fluid
	24	780	albite + fluid
4455†‡	8	820	albite + liquid + fluid
	24	780	albite + liquid + fluid
4456‡	8	800	albite + liquid + fluid
	24	760	albite + fluid
4457‡	48	780	albite + liquid + fluid

†The bracketing runs were held at the high T for the indicated time and then rapidly cooled to the low T.

‡These runs are hypothetical; they were constructed to illustrate points made in the text.

bracket may be needed to test the thermodynamic model. In this case it would be necessary to make run 4457 and increase the run duration to 48 hours to be on the safe side. The result is melting, so one can say with certainty that, in this hypothetical case, melting occurs between 760 and 780°C.

Note that the above case is for a very simple reaction. Thinking out ways of demonstrating reversals of more complex reactions is more difficult. Many published works rely on another technique to demonstrate attainment of equilibrium, that of doing the same experiment for increasingly long times and showing that the phase assemblage and phase compositions do not change with time. This is a necessary, but not a sufficient, condition of equilibrium. For example, one could have run the 790°C experiment for 500 hours and compared it to the result of run 382. Even if the crystals showed no signs of melting, it still has not been proved that melting would not occur eventually.

In many systems, reaction rates become very sluggish near the true equilibrium boundary. This is because the rate of reaction is approximately proportional to the free-energy driving force (ΔG) of the reaction and ΔG goes to zero at equilibrium. The reaction:

$$\text{pyrophyllite} \rightleftharpoons \text{diaspore} + \text{andalusite} + \text{water} \qquad (2.1)$$

was studied by Haas & Holdaway (1973) using single crystals of andalusite to monitor reaction direction. For experiments of constant time the single crystal of andalusite shows increasing weight loss with decreasing temperature below the equilibrium boundary (Fig. 2.2). In general, these sluggish reactions show a temperature interval of little or no reaction bracketing the equilibrium boundary. The point of equilibrium must lie somewhere within that temperature interval but not necessarily in the middle of it. In fact it is sometimes possible to argue that the true position of equilibrium lies well toward one side of the "bracket." This would be the case in the albite melting experiments above, because experience has shown that less T overstep is necessary to melt crystals than to crystallize liquids, especially viscous melts such as feldspar compositions.

Before leaving the discussion of reversals, we want to point out that it is entirely possible to reverse a metastable equilibrium. The technique has often been used successfully and will be discussed along with other examples of reversals in later chapters.

There is another system-specific problem waiting to trap the unwary. The general effect is to change the bulk composition of the system during the run. In an invariant system it would not matter, but few experiments are done on invariant systems and, for the rest, changing system composition changes phase compositions, which in turn change phase stabilities and ruin the experiment. There are three subsets of this phenomenon of which we are

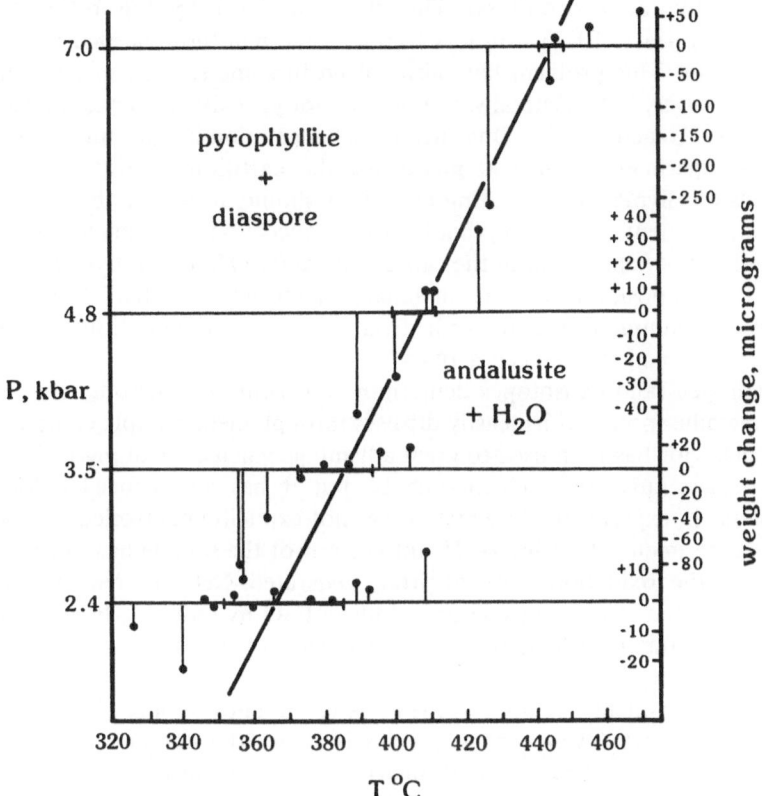

Figure 2.2 *P–T* diagram for Reaction (2.1). The position of the equilibrium boundary (heavy diagonal line) is based on changes in weight of andalusite single crystals. Negative weight changes indicate andalusite is dissolving, positive changes that it is precipitating. The light vertical lines connect initial and final weights. Note that the weight change increases as the temperature difference between the run and the equilibrium value increases. Redrawn from Haas and Holdaway (1973).

aware (and almost certainly some we have not yet encountered). Those three are:

(a) Reaction between the system and its container;
(b) The loss, or gain, of a component through the sample container walls;
(c) Solution of substantial amounts of silicate components in a fluid phase.

These problems are only important in certain systems; unfortunately those systems are some of the geologically most important.

Iron loss to the sample container is by far the most common reaction between sample and container. This problem is most serious in mafic and ultramafic systems which are iron-rich and which melt at high *T*(high *T* enhances the rate of iron loss). The effects can be staggering in the worst cases, with nearly 100% iron loss in times less than those needed to reach equilibrium. This problem has been solved in some specific experimental configurations, but a general solution does not yet exist. Other examples of this type of reaction exist. One such is nickel solution in platinum capsules during experiments aimed at measuring the partitioning of Ni between crystals of olivine and coexisting melt. One should anticipate this problem with any element, such as Fe, that is easily reduced to the metal and that has a small activity coefficient in the capsule metal (see Hultgren *et al.* 1973 for activity–composition relations in binary metal alloys). Alkali loss, especially of sodium, can also occur from melts under conditions of high temperature and low oxygen activity.

Hydrogen (and its isotopes deuterium and tritium) is the only common non-metallic species which easily diffuses through metal sample containers. This behavior has been used to great advantage when controlling hydrogen and oxygen fugacities in experiments, but it has also caused a lot of problems in experiments in which it was not explicitly controlled. In these latter experiments diffusion of H_2 into or out of the sample has drastically changed the oxidation state of iron, generated CH_4 by reacting with graphite in the sample capsule or produced H_2O by reducing Fe_2O_3 in the sample. By far the best solution to this problem is external control of H_2 fugacity.

Many experiments are done in the presence of a nearly pure H_2O fluid. If the fluid : silicate ("water : rock") ratio is large, the composition of the condensed phases (crystals and melt) may be changed by differential dissolution. A common example is the preferential solution of silica and alkalies from feldspar compositions, leaving a condensed phase composition enriched in alumina, which is very insoluble in aqueous fluids. In experiments in which the aqueous solution does not require analysis, this problem is easily solved by keeping the H_2O : silicate ratio just high enough to ensure the presence of a fluid phase. If large fluid : silicate ratios must be used in order to have enough fluid for analysis, the composition of the starting material must be adjusted to allow for the differential solution.

2.5 What comes out: analytical techniques

Let us assume that all experimental problems have been circumvented and that a "good" run has been completed. The next task is to describe the products as accurately as possible. For crystals and glass the approach is much like a modern description of a rock, and the tools used are the same. In this section we present an overview and describe some problems unique to examining experiments. Techniques for analyzing fluids from experiments are described in Chapters 5 and 8.

It is necessary to identify the phases present, determine their elemental compositions, their structural states, and finally, determine if they were present at run conditions or are "quench" phases formed as the run cooled from run T to room T. A useful sample for these purposes is a polished thin section. It can be used for examination with an optical microscope as well as for analysis by several microbeam techniques. The microbeam instruments which measure element composition include the electron microprobe (EMP), scanning electron microscope (SEM) equipped with energy-dispersive X-ray analyzer (EDS), and the ion microprobe. Of these the SEM is the easiest to use and most useful because its high magnification capacity allows observation of many textural features in run products that are below the resolution of optical microscopy. The newest electron microprobes generate excellent back-scattered electron images comparable to those obtained by SEMs. The ion microprobe is a powerful new analytical tool for measuring the light elements H to Na which are difficult or impossible to determine with an EMP. Ion probes can also be used to measure trace element and isotope abundances in favorable cases.

Elemental analysis of individual phases is a powerful technique for evaluating the quality of an experiment, for instance by searching for compositional zonation or by testing for sensible major-element partitioning relations. Zoned crystals are clear evidence of lack of equilibrium, as are partition coefficients that violate known values. A commonly used relationship is the partitioning of FeO and MgO between olivine and melt. Roeder & Emslie (1970) found the partition coefficient (in moles):

$$K_D = \frac{FeO_{Ol} \times MgO_m}{MgO_{Ol} \times FeO_m} = 0.30 \pm 0.03$$

This relationship has been confirmed in many experiments in which most of the iron was in the $+2$ oxidation state. Thus, if coexisting olivine and glass from a given run yield a significantly different value (usually higher), it is strong evidence that the two phases were not in equilibrium. If the olivine has a higher Fe : Mg ratio than predicted, it suggests that either the olivine is a quench phase (see below) or that it was present during the run, but that iron was lost from the melt at a higher rate than from the already crystallized olivine (because of higher diffusion rates in the melt). Another good test of the run and of the analytical techniques is to calculate the mode of the run using elemental analyses of all of the coexisting phases and of the bulk composition of the starting material. The calculated mode should then be checked by comparison with that measured by point counting or by checking its mass balance (if the equilibrium has a variance greater than 2).

Characterizing the details of run product structure can be important in unraveling thermodynamic behavior. This is especially true for phases showing strong order–disorder effects. Until recently the means of characterizing structure required working with bulk samples. These methods are X-ray diffraction, Mossbauer spectroscopy, and high-resolution nuclear magnetic resonance (NMR) spectroscopy. The situation has been dramatically changed by the microbeam techniques of transmission electron microscopy (TEM), micro-Raman and micro-infrared (IR) spectroscopy. These techniques allow determination of structural characteristics on the micrometer to sub-micrometer scale, which is the minimum size of crystals in run products. Micro-Raman spectroscopy also allows some characterization of glasses.

Run products often contain quench phases: phases not present during the run, but which formed during the quench from high P and T to room T and atmospheric P. These phases may either be crystals which form from the melt, or crystals or glass which form from the fluid phase. Some quench crystals have distinctive textures and are readily identifiable (Wyllie 1963), but this is not always the case, and sometimes they can only be identified from their chemical compositions. Quench crystals may be separate entities, or they may form rims on existing crystals. A common example of the latter is formation of thin (1–5 μm) rims of iron-rich quench olivine on stable olivine crystals. Even though the rims are thin, they may constitute a significant volume of the system. These quench crystals can cause a large change in the composition of the melt, so that the composition of the quenched glass may be quite different from the melt which was actually present at the P and T of the experiment (Green 1976).

One of the most powerful tools for inspecting run products is the polarizing microscope. It should always be used prior to the microbeam techniques, and in some cases it is far superior for phase identification. A case in point is identification of glass to determine the solidus T in melting experiments. In complex phase assemblages the amount of melt formed at the solidus may be very small, for example in H_2O-saturated melting of basaltic compositions. The most sensitive technique for determining the presence of glass is to make a grain mount of a finely ground portion of the run product. Using an index oil close to the average value of the crystals, the glass will stand out dramatically due to its low refractive index compared to the crystals.

2.6 Conclusions

In conclusion, a good experiment involves five carefully considered steps:

(1) *Experimental design*. The goal of the experiment should be well formulated, for example the testing of a thermodynamic model.
(2) *Starting material*. The starting material should be selected to enable

equilibrium to be demonstrated unequivocally. For example, a sub-solidus reaction would generally contain crystalline materials of both reactants and products.

(3) *Control of P, T, and composition.* P and T need to be controlled at levels needed for testing the model under consideration. Compositional changes can occur in many systems and detailed analyses of interactions with capsule materials, furnace gases, etc. need to be made.

(4) *Proof of equilibrium.* In general a true reversal is required.

(5) *Analysis of products.* Optical microscopy is the first and easiest technique to be used. The use of other techniques is dictated by the experiment being done.

3 The machines

3.1 Introduction

This is the "nuts and bolts" chapter where we set out brief descriptions of the experimental devices used in a wide range of situations. We have covered most of the commonly used types and some interesting ones seldom used. The descriptions are not meant to be exhaustive; references are given for more detailed information. The Appendix contains elaborations of some of the points mentioned in this chapter. Each description follows a common format for easy comparison.

The machines fall into different types mainly dependent on the $P-T$ range for which they are designed. Those ranges are shown in Figure 3.1. An important fact of life in the experimental world is the inverse relation between a machine's upper pressure limit and its sample volume. This is shown schematically in Figure 3.2 in which sample capsule sizes are shown for many common machines. The larger the sample volume, the greater the flexibility possible in the experiment and the more easily and precisely intensive variables such as P, T, and oxygen fugacity (f_{O_2}) can be controlled. Another major difference between machines is the quench rate; this is the rate at which the sample temperature decreases at the termination of the run. Each of these factors is mentioned in the following descriptions.

3.2 Machines for "easy" conditions (atmospheric pressure)

These machines can be very simple in design if their only purpose is to provide high temperatures in air. These "quench-furnaces" were used by J. Frank Schairer at the Geophysical Laboratory to unravel the melting relations of simple, iron-free systems. The complexity of the apparatus increases with the addition of a controlled oxygen fugacity capability.

Machine:

Quench and gas mixing furnace

	P	T
Range:	Vacuum to 1 bar	Up to 1,600°C
Precision:	About 1%	1 to 15°C
Accuracy:	1%	1 to 30°C

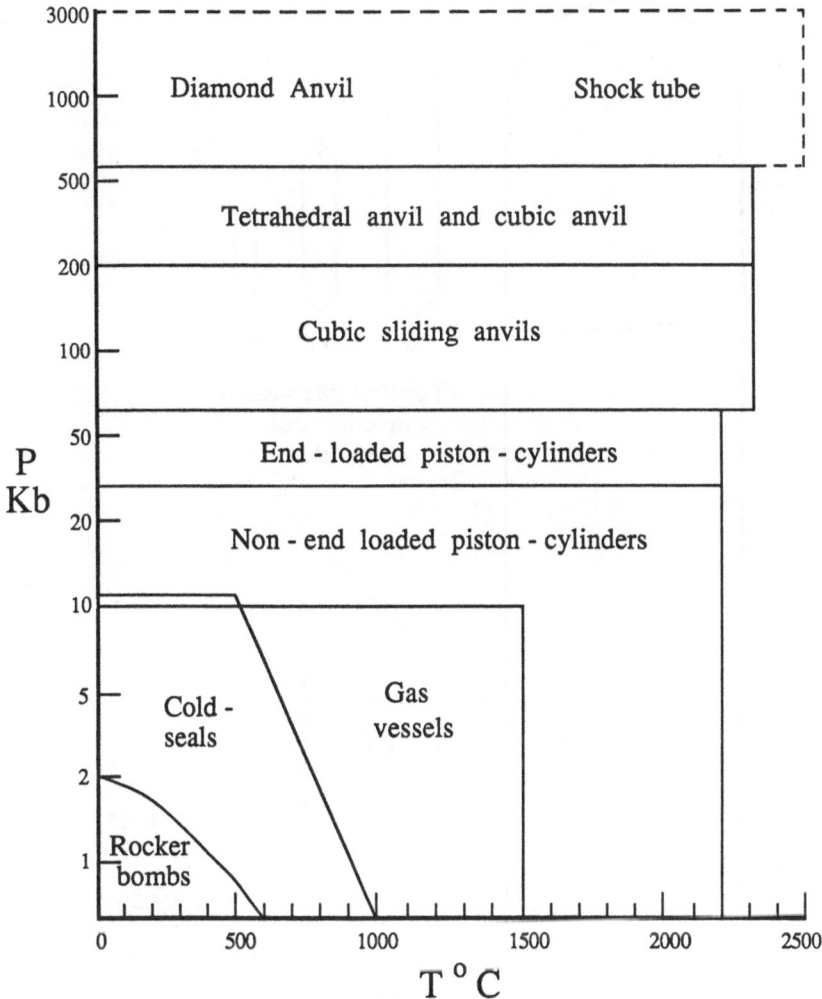

Figure 3.1 *P–T* ranges of pressure machines.

Sample volume: Usually small, about 0.1 ml, but up to 50 ml possible.

Original references: Darken & Gurrey (1945) original design, Osborn (1959) application to magmas.

Working references: Biggar (1974) calibration techniques, Nafziger *et al.* (1971) description of the technique. Donaldson (1979) Na and Fe loss during experiments. Huebner (1987) provides up-to-date summary.

Advantages: Precise control of *T* and f_{O_2}, rapid quench rate, nearly zero

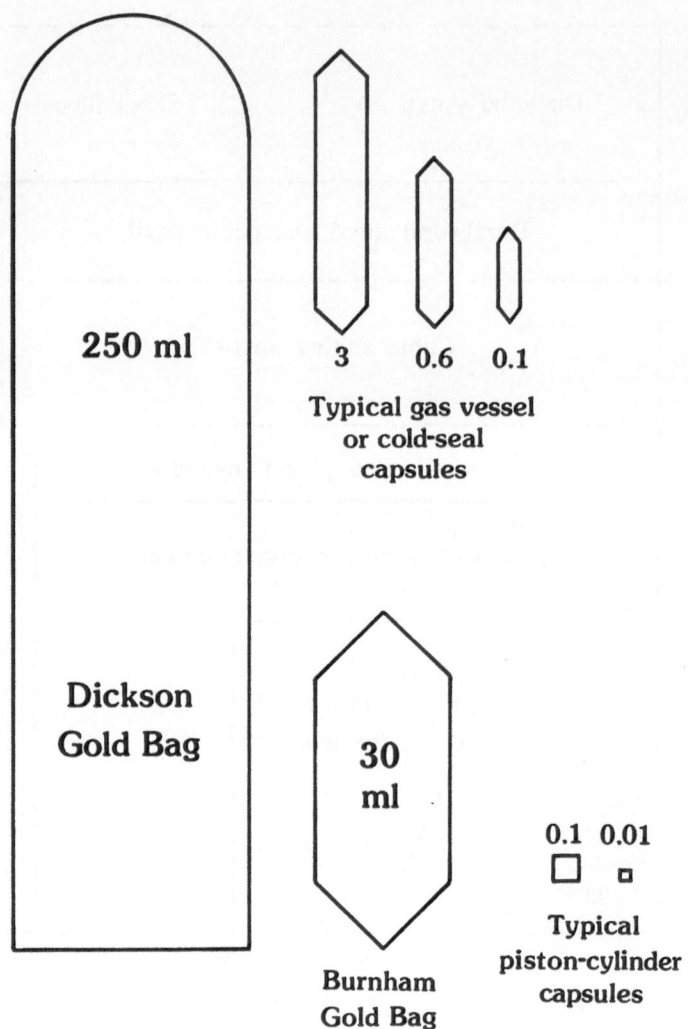

Figure 3.2 Sizes of typical sample capsules. The approximate useful volume is shown in ml.

iron loss, multiple samples per run. Easy to use, can be accurately calibrated for T and f_{O2}.

Disadvantages: A major problem is sodium volatilization at $T > 1,100°C$ when f_{O_2} is lower than fixed by the quartz + fayalite + magnetite (QFM) assemblage.

Typical applications: Liquidus relations in iron-bearing synthetic systems and rocks, synthesis of iron-bearing minerals and glasses, element partitioning experiments.

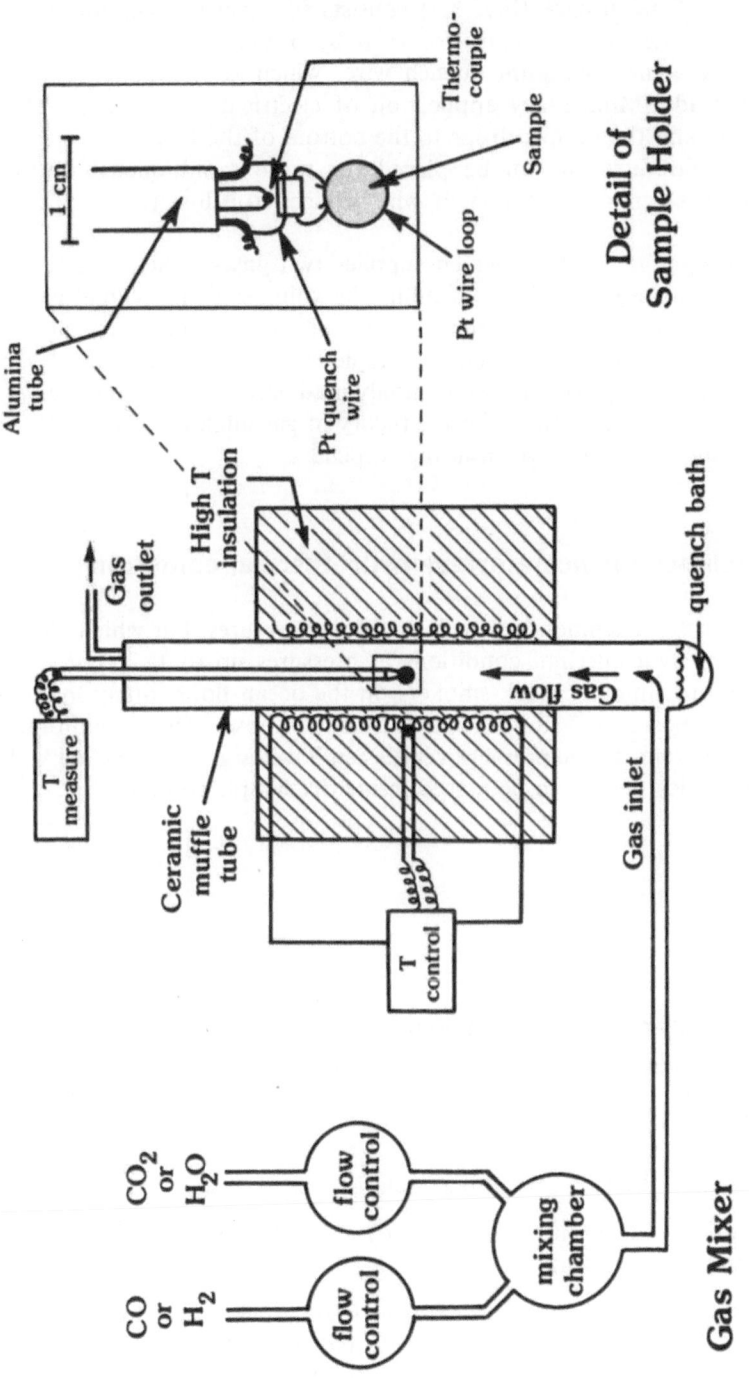

Figure 3.3 Schematic diagram of a quench furnace apparatus. The detail on right shows the sample and quench loop arrangement. Control of f_{O_2} by gas mixing is described in the Appendix.

Description: The furnace (Fig. 3.3) consists of a vertical ceramic tube surrounded by heating elements, followed by insulation. The sample is suspended on a fine platinum quench wire, which is connected to two thicker Pt leads. Momentary application of electrical current ·melts the quench-wire, and the sample drops to the bottom of the furnace in which water or liquid nitrogen can be placed for more rapid quench rates. Suspending the sample on a fine Pt wire reduces iron loss to acceptable values.

The gas mixer is a system which supplies two gases mixed in known ratios. The ratios are fixed by controlling the volumes of individual gases which then pass to the mixing chamber. The mixed gases flow through the furnace tube and fix the f_{O_2} around the sample so that the oxidation state of iron can be controlled. The most commonly used gases are CO and CO_2 or H_2 and CO_2. See the Appendix for the theory of gas mixing. Details of gas mixing techniques are also given in the Appendix.

3.3 Machines for near-surface hydrothermal conditions

We now consider machines for much lower temperatures, but which allow simulation of hydrothermal conditions at pressures up to 1–2 kbar, i.e. deep enough to simulate black-smokers on the ocean floor, many hydro-thermal ore deposits, and sites for nuclear waste disposal. These conditions are quite moderate, so the pressure vessels can have large volumes and will allow some fancy devices to be placed inside the sample container.

Machine:

Dickson rocking autoclave

	P	*T*
Range:	1–2000 bar	Up to 600°C
Precision:	0.02%	±10°C
Accuracy:	0.1%	±10°C

Sample volume: Usually 250 ml but could be made larger.

Original references: Dickson *et al.* (1963) original design, Seyfried *et al.* (1979) currently used design.

Working references: Bischoff & Dickson (1975) rock–water interaction. Jenkins *et al.* (1984) nuclear waste evaluation and new techniques. Bischoff & Seyfried (1978) seawater. Seyfried *et al.* (1987) detailed description.

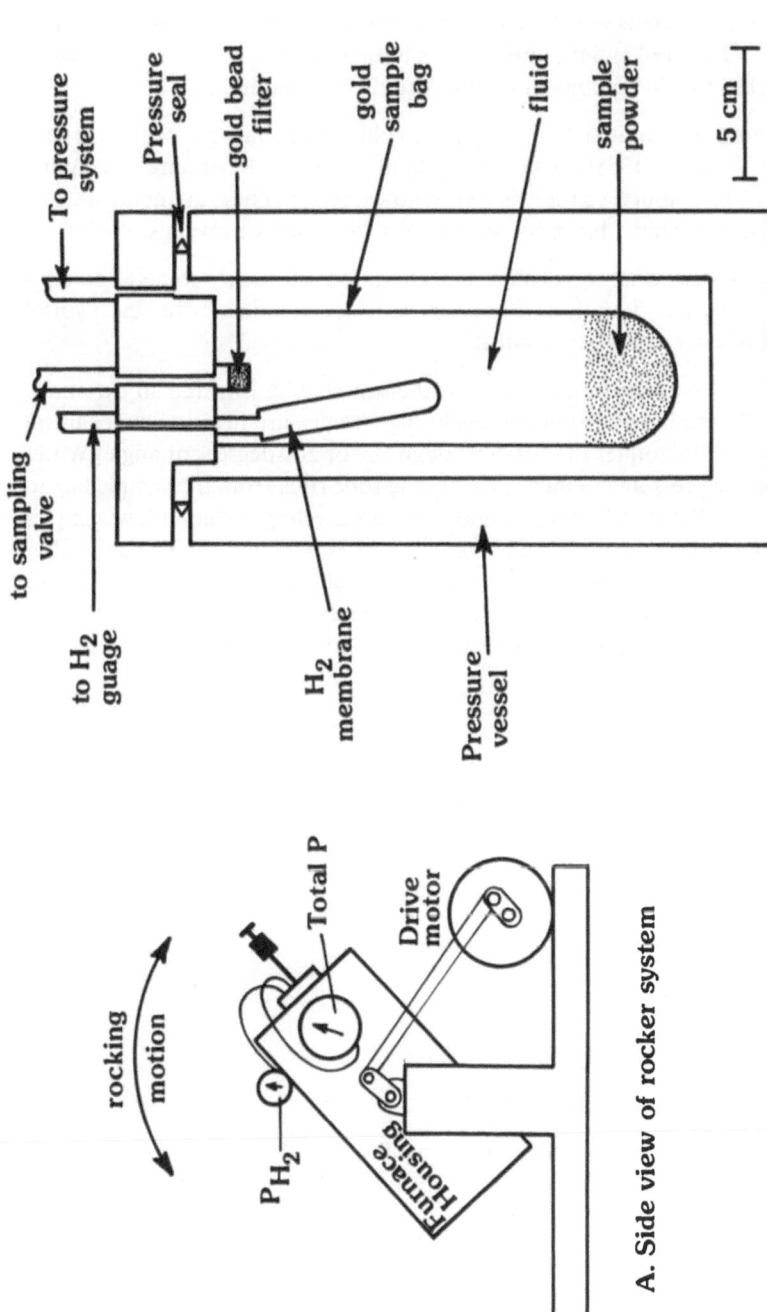

Figure 3.4 Dickson rocking autoclave. (A) Side view of the rocking mechanism. (B) Cross-section of the vessel, gold bag, H_2 membrane, and sampling tube with gold filter.

B. Cross-section of vessel and sample bag system

To pressure system

Pressure seal

gold bead filter

gold sample bag

fluid

sample powder

5 cm

to sampling valve

to H_2 guage

H_2 membrane

Pressure vessel

A. Side view of rocker system

rocking motion

Total P

Drive motor

P_{H_2}

Furnace Housing

Advantages: Large sample volume, hydrostatic pressure, direct sampling of fluid at P and T. The rocking motion stirs the sample which results in increased reaction rates. There is lots of room inside the sample bag for sensors for Eh, pH and H_2. The gold and titanium system is inert to most fluids. This system is not restricted to $P-T$ points along the liquid/vapor boundary.

Disadvantages: Limited $P-T$ range, the gold sample bags are expensive [but Seyfried & Janecky (1985) are now using titanium bags]. Beginning and ending runs takes 1 to 3 hours, so there is a very slow quench. These are good systems for the study of fluids, but poor systems for the study of crystals.

Typical applications: Hydrothermal alteration of rocks, mineral solubilities, hydrothermal alteration of nuclear waste forms, kinetic studies. Typical run durations are weeks to months.

Description: As shown in Figure 3.4, the sample bag is inside an externally heated steel vessel. The vessel and enclosing furnace are held in a mechanism which continually rotates the vessel through 180 or 360 degrees of angle. Water is used for the pressure medium. A sampling tube runs from the sample bag to the outside of the vessel. A valve and sample collecting system allow samples of the fluid to be taken while the sample is at run P and T. The gold filter on the inside end of the sampling tube prevents particulate samples from contaminating the sampled fluid. The H_2 membrane can be used to either control or measure the H_2 pressure and thus, indirectly, oxygen fugacity.

Machine:

Barnes rocking autoclave

	P	T
Range:	1–2000 bar	Up to 600°C
Precision:	0.02%	±1°C
Accuracy:	0.1%	±2°C

Sample volume: Usually one liter, fixed.

Original references: Barnes (1963)

Working references: Barnes (1971) and Bourcier & Barnes (1987) current description. Rimstidt & Barnes (1980) for an application to kinetic studies.

Advantages: Large sample volume, hydrostatic P, direct sampling of fluid at P and T. The rocking motion stirs the sample which increases reaction rates. The fixed volume makes this a particularly good system for studying two phase, liquid + vapor systems.

Disadvantages: The fixed volume makes this system more difficult to use for studies in one-phase, supercritical fluid systems than is the Dickson apparatus. It is also more difficult to make the sample container inert to reactive fluids.

Typical applications: Has been used extensively for studies in aqueous sulfide systems, both for measuring sulfide mineral solubilities and to determine speciation in the fluid.

Description: The system is very similar to the Dickson system except that there is no sample bag inside the vessel. Therefore the inside of the vessel forms the sample container, and the sample fluid is the pressure medium. The sample valve arrangement is also more complex than is commonly used with the Dickson apparatus, in order to allow sampling of both the liquid and vapor phases.

3.4   Moderate pressure: crust–upper-mantle machines

We now move to the range of metamorphic and igneous petrology experiments. The three machines described here are the ones which have been used to dramatically increase our understanding of metamorphic and igneous processes. All three were initially developed at the Geophysical Laboratory of the Carnegie Institute of Washington between 1945 and 1960. We discuss them in order of their upper pressure limit.

Machine:

Cold-seal vessel ("Tuttle bomb")

	P	*T*
Range:	1–12,000 bar	Up to 950°C
Precision:	0.1 to 1%	±1 to ±10°C
Accuracy:	0.1 to 1%	±1 to ±30°C

Sample volume: Typically 0.1 to 0.5 ml in standard vessel but can be considerably larger.

Original references: Tuttle (1949) for the 0–3 kbar version. Luth & Tuttle (1963) for the 12 kbar version.

Working references: Boettcher & Kerrick (1971) describe temperature gradients and how to minimize them. Williams (1968) describes a high temperature type for operation to 1200°C. Rudert *et al.* (1976) describe the rapid-quench type. Kerrick (1987) gives a recent description.

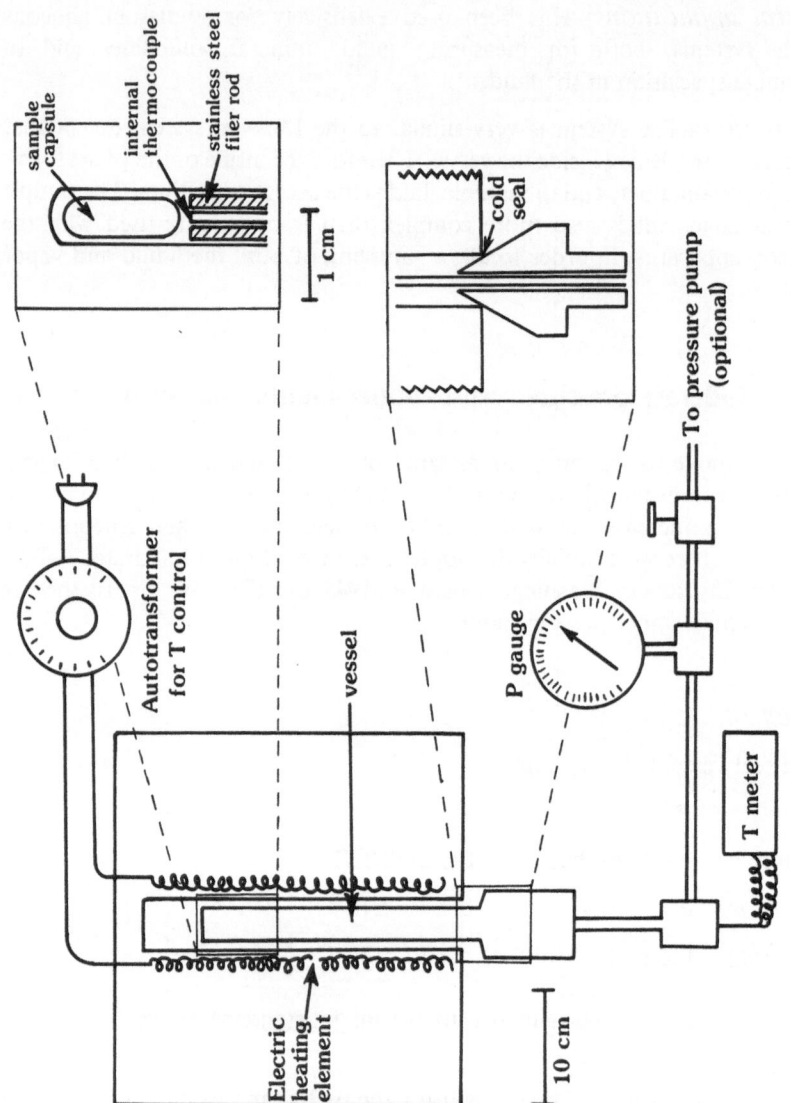

Figure 3.5 Cold-seal pressure system.

Advantages: These are the simplest and least expensive of the common high *P* systems. Easy to use, true hydrostatic pressure. Internal thermocouples allow accurate *T* measurement.

Disadvantages: External heating results in an inverse relationship between operating *T* and *P*, and severely limits the working *T* range.

Typical applications: This is the machine of Tuttle & Bowen's (1958) classical experiments on melting of granites. It has also been the workhorse of metamorphic experiments, and is also used extensively for hydrothermal synthesis of minerals. The rapid-quench model can be used for kinetic experiments such as diffusion, as well as for studies of fluid phase chemistry.

Description: See Figure 3.5. The vessel is a rod of a superalloy, usually Stellite-25 or Rene-41 (but better alloys are now available, see Appendix), with an 0.64 cm hole drilled almost through to make a test-tube shape. The end of the vessel with the pressure seal is outside the furnace, hence "cold-seal," and is formed by a simple cone seal (see inset in Fig. 3.5). The pressure medium is usually water, but argon must be used above 8 kbar pressure. Furnaces typically have Kanthal windings and *T* can be controlled with simple, inexpensive systems.

Machine:

Internally heated gas vessel

	P	*T*
Range:	0.1 to 10 kbar	To 1,500°C
Precision:	0.1%	±2°C
Accuracy:	0.1%	±5°C

Sample volume: Typically 0.6 to 4 ml, can be up to 30 ml.

Original references: Yoder (1950) small volume version, Burnham *et al.* (1969b), large volume version.

Working references: Holloway (1971) provides a general description. Ford (1972) describes a modified small volume version. See Lofgren (1987) for a recent update.

Advantages: *P* and *T* limits are not inversely proportional; true hydrostatic *P*; accurate *P* and *T* measurement; well suited to use of hydrogen membrane to control f_{O_2}; large sample volume. Room inside vessel for "exotic" apparatus. The *T* gradients over the sample can be kept very small.

Disadvantages: Relatively complicated and expensive pressurizing system.

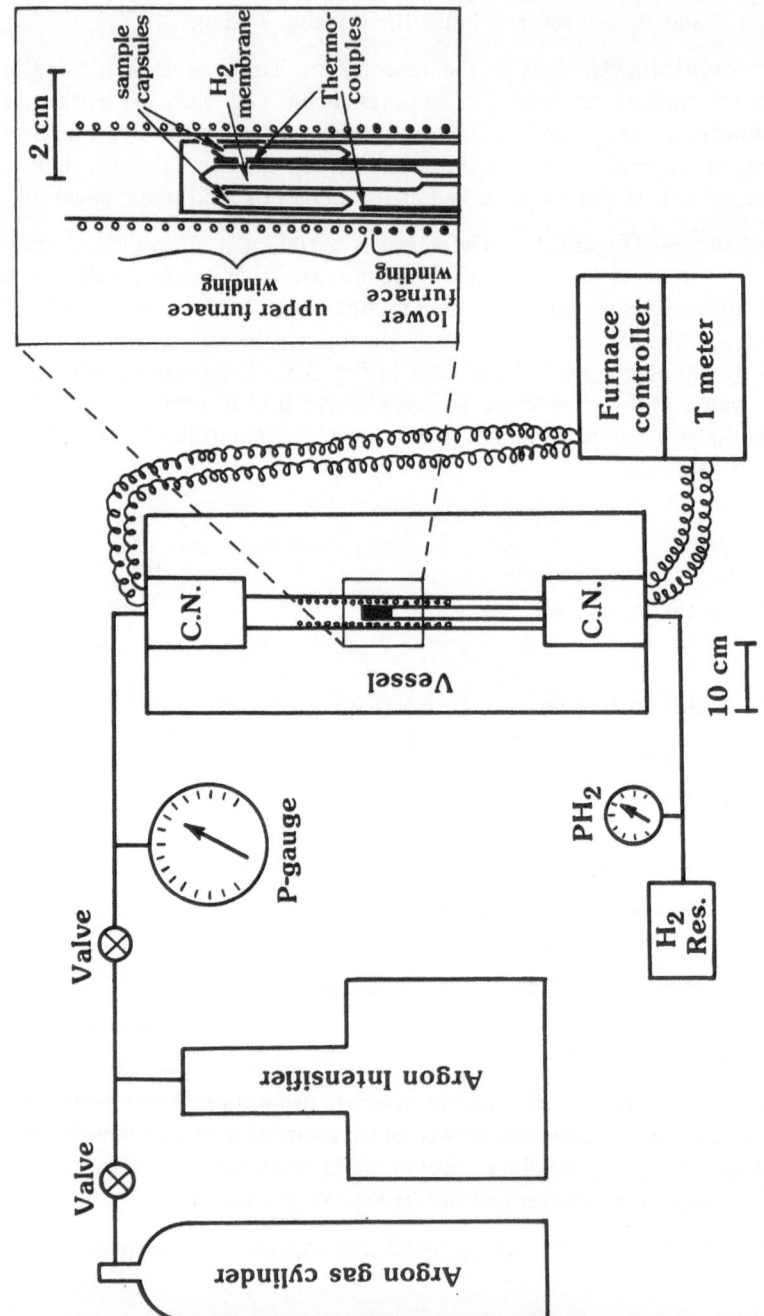

Figure 3.6 Internally heated, argon medium high pressure vessel and pumping unit. C.N. = closure nut. The vessel is shown equipped with a Shaw H₂ membrane and H₂ control system.

Furnaces are difficult to make, especially for $T > 1,200°C$. T control systems require more sophistication than the ones used for cold-seals. Relatively slow quench rate. Can be subject to large T gradients over the sample volume.

Typical applications: Ideal for melting relations in hydrous andesitic to rhyolitic compositions. Also the best choice for high T fluid/melt or fluid/crystal equilibria where a large fluid volume is needed for analysis. Very useful for synthesizing large volumes of starting material.

Description: These vessels are distinguished by the placement of the furnace inside the vessel: hence the name "internally heated," in contrast to cold-seal vessels which are externally heated (Fig. 3.6). The pumping system shown is highly schematic. The vessel shown has two independent furnace elements and is independent for operation in a vertical position.

Pumping system: Uses a large area hydraulic piston to drive a small area piston to compress the argon gas pressure medium. Two or three stages of compression are used to generate 10 kbar. A typical system would pump from cylinder pressure, 100 bar, to 500 bar in the first stage, 3 kbar in the second stage and to 10 kbar in the third stage.

Vessel: The vessel is a thick-walled steel cylinder having one or both ends open. The open ends are closed by heads through which pressure, electrical and thermocouple leads enter.

Furnace: The electric furnace consists of one or two coiled wires surrounded by high temperature insulation. The wire elements are made of Fe–Ni–Cr alloys for use to $1,100–1,200°C$ or Pt–Rh for higher T. Argon gas has a high density at high P, for instance at 5 kbar and $1,000°C$ its density is about that of room temperature water. Inside the furnace strong convective currents are set up in the dense argon. The convection can cause very large T gradients for vessels operated with their long axis horizontal. Double element furnaces operated vertically (as shown) use convection to eliminate gradients. Such furnaces can achieve isothermal zones with lengths of 6 cm. As shown in the inset to Figure 3.6, it is common to equip internally heated vessels with H_2 membranes for control of oxygen fugacity (Gunter *et al.* 1987).

Machine:

Piston-cylinder

	P	T
Range:	5–60 kbar	Up to $1,800°C$
Precision:	± 0.1 to ± 3	± 3 to $\pm 10°C$
Accuracy:	± 0.1 to ± 5	± 3 to $\pm 25°C$

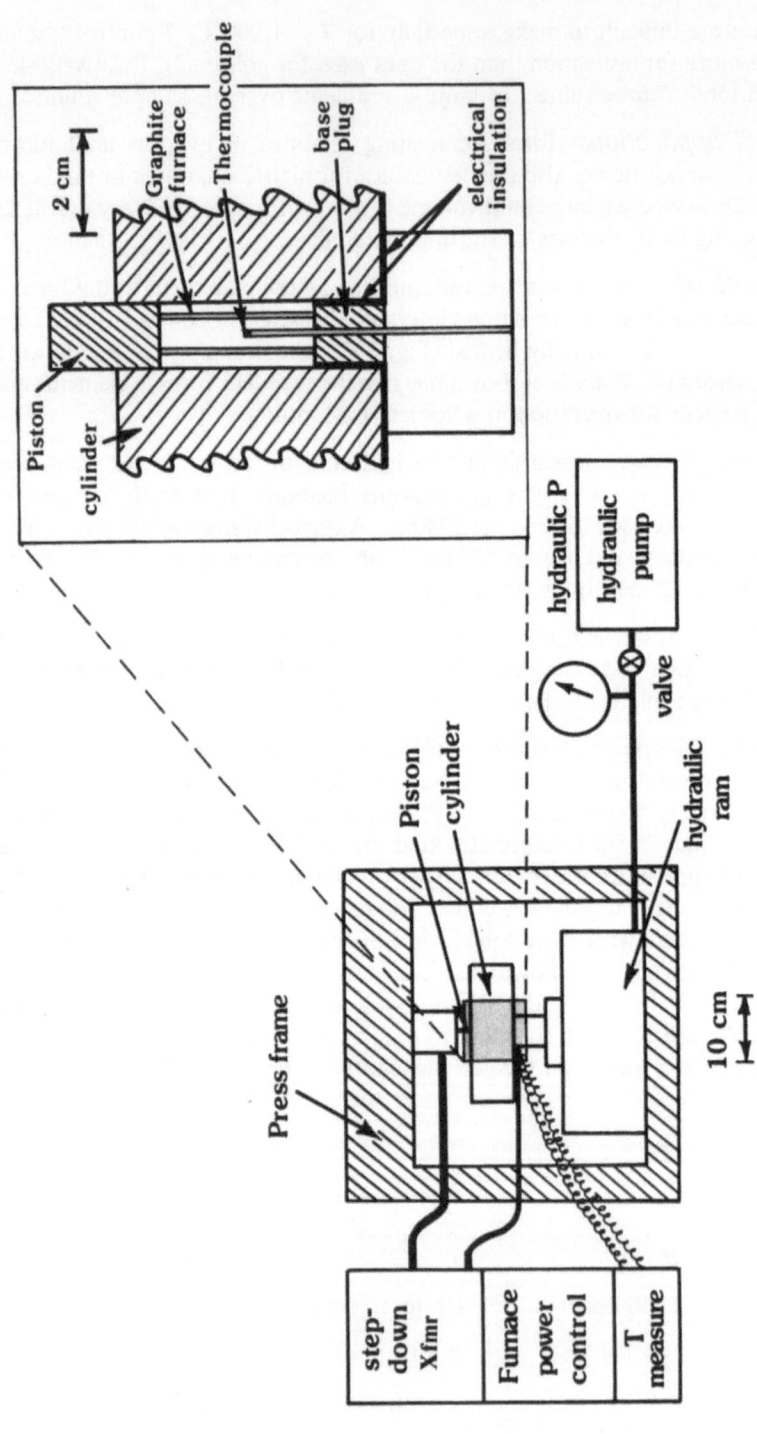

Figure 3.7 Piston–cylinder high pressure system. Detail at right shows the pressure chamber in cross-section.

30

Sample volume: Typically 0.01 to 0.1 ml

Original references: Boyd & England (1960).

Working references: Johannes *et al.* (1971) discusses interlaboratory *P* calibration. Mirwald *et al.* (1975) and Boettcher *et al.* (1981) discuss low friction furnace assembly designs. Bohlen (1984) describes recent *P* calibration techniques.

Advantages: Very easy to operate, especially up to 30 kbar and 1,700°C. Very rapid quench rate.

Disadvantages: Limited sample volume, large *T* gradients, requires attention to *P* and *T* calibration to avoid large uncertainties. The *P* effect on thermocouples makes high accuracy at high *P* and *T* difficult to obtain. Systems for *P* > 30 kbar are expensive.

Typical applications: This is the work-horse upper mantle machine. Its use has provided most of our understanding of melting relations in basaltic and ultramafic systems. Careful phase equilibria studies in simple systems have provided much of our knowledge of the thermodynamic properties of pyroxenes and garnets.

Description: Consists of a simple piston pressing into a cylinder (Fig. 3.7). Pressure is generated by forcing a piston into a cylinder and compressing the solid materials in the furnace assembly. Pressure is calculated by measuring the *P* on the hydraulic ram and then multiplying by the ratio of ram area to piston area. Depending on the area of materials used in the furnace assembly, there may be a large correction necessary to adjust the calculated pressure. In any

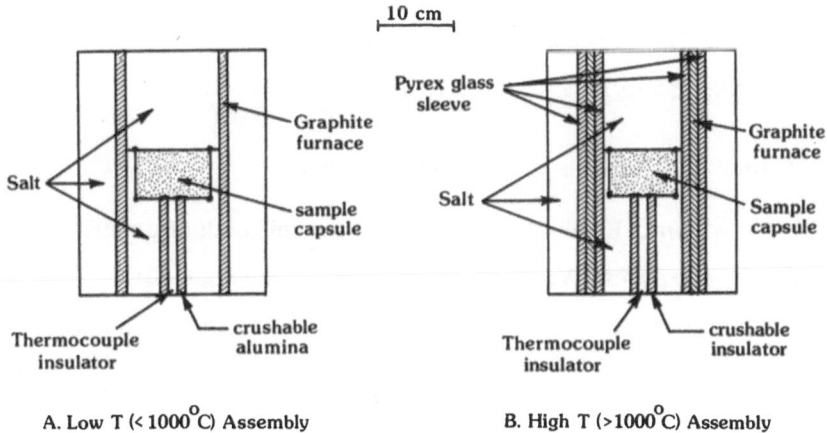

Figure 3.8 Piston-cylinder furnace assemblies.

good paper describing piston-cylinder results there will be a discussion of pressure calibration.

Temperature is generated by passing a very high current through the graphite tube furnace. The furnace is inside the vessel (and thus is internally heated). The entire furnace assembly is expended in each run. Many types of furnace assembly have been used. Two are shown in Figure 3.8. One is designed for minimum friction to provide the best P accuracy and should be used whenever possible. Unfortunately it has an upper T limit of 1,000 to 1,100°C. At higher T the NaCl parts in the hot-spot melt and the graphite furnace disintegrates. More refractory materials therefore become necessary. The best choice is probably glass. Borosilicate glass (Pyrex) which softens at about 700°C is commonly used.

3.5 Ultra-pressure machines

We have now covered the common machines in which experiments can be done in anything like a routine manner. The devices we are about to discuss are still plagued by large uncertainties in P and T, by the inability even to measure T, and generally by very small sample sizes. These machines are in the process of extensive development and will be improved dramatically in the next decade. They allow us a glimpse of the deep earth – the mid to lower mantle – and even the core.

Machine

Diamond anvil

	P	T
Range:	To >2,000 kbar	Up to 3,000°C
Precision:	±5 to 10%	±10°C at $T < 1,000$°C ±100°C at $T > 1,000$°C
Accuracy:	±5 to 10%	as for Precision

Sample volume: Extremely small, around one millionth of a milliliter.

Original references: Weir et al. (1959), van Valkenburg (1963).

Working references: Bassett (1979) general review and description. Xu et al. (1986) describe the latest ultrapressure achievement. Heinz & Jeanloz (1984) describe pressure standards. Jephcoat et al. (1987) describe operation. Mineral solubility measurements are described in van Valkenberg et al. (1987).

Advantages: Capable of generating very high P in a very small apparatus for

32

a very low cost. The transparent diamond windows and small size permits use of the device under a polarizing microscope, in an X-ray diffractometer, or in various spectrometers. In these systems phase changes can be viewed directly.

Disadvantages: Although very easy to use at room T, it is difficult to control higher temperatures. It is very difficult to generate T above $900°C$ in a controlled manner. The very small sample volume is a limitation. Subject to extremely large P gradients across the face of the diamond anvils.

Typical applications: Exploration of mid to lower mantle phase equilibria. Measurement of mineral compressibilities and thermal expansions at high P. Exploration of the terrestrial core/mantle boundary. Studies of phase changes in the interiors of Jupiter and Saturn.

Description: Two diamond anvils with a thin metal gasket and a layer of sample between them are pressed together in a small frame (Fig. 3.9). Force multiplication is achieved by levers and is applied by spring-loading. P multiplication is achieved by very small areas on the diamond faces. In general the higher the P sought, the smaller the diamond face. The metal gasket is used to contain the sample and reduce the P gradient. Even fluids can be studied using such a gasket. P is determined by measuring the change in cell volume in gold or NaCl (by X-ray diffraction) or by measuring the change in wavelength for ruby fluorescence, or by dividing the applied force by the measured area of the diamond face. Heating below about $900°C$ can be done with an external

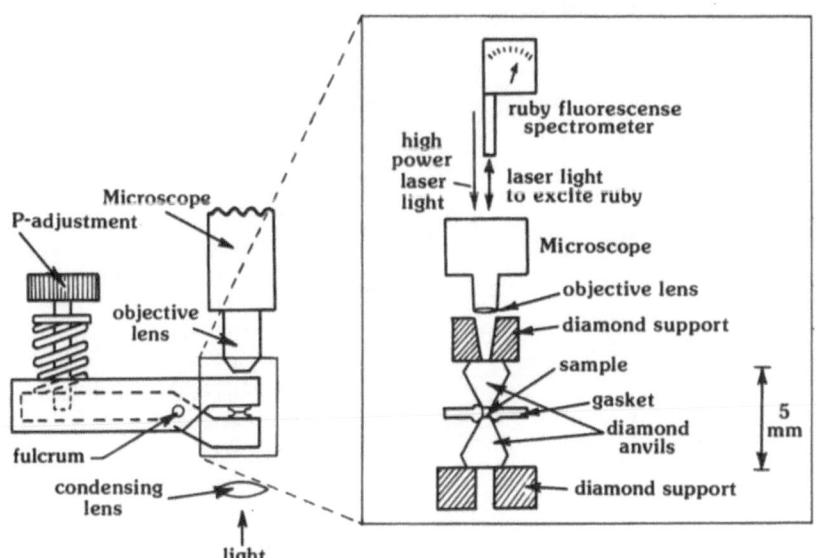

Figure 3.9 Diamond anvil cell.

furnace. Laser heating is used to generate higher T. This works well for the limited area of the laser beam (about 10 μm) but it is difficult to measure T accurately.

Machine:

Multiple anvil device

	P	T
Range:	Up to 300 kbar	Up to 3,000°C
Precision:	±10%	±10 to 30°C at $T < 2,000$°C
Accuracy:	±10%	±10 to 100°C at $T < 2,000$°C

Sample volume: 0.001 to 0.1 ml

Original reference: Hall (1960)

Working references: Papers in Akimoto & Manghnani (1982), summary by Liebermann *et al.* (1985).

Advantages: Much larger sample volume and much better T control than with the diamond anvil. With a synchrotron X-ray source unit cell parameters and phase transitions may be determined at P and T.

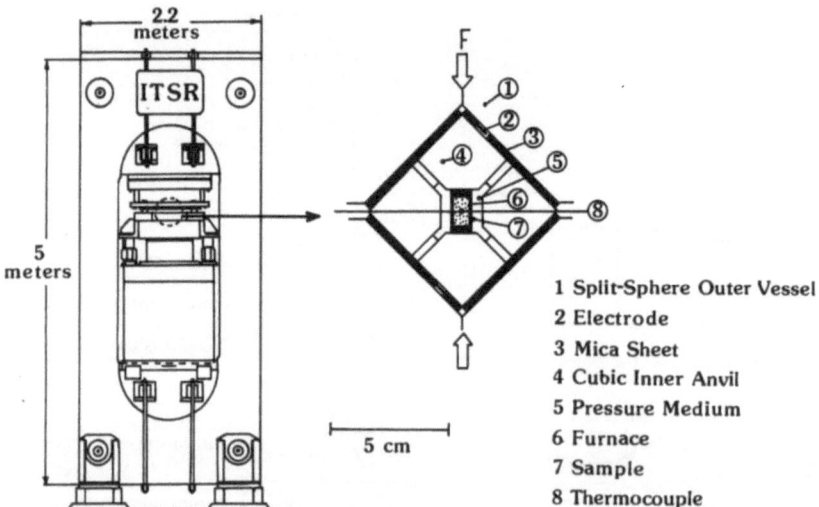

1 Split-Sphere Outer Vessel
2 Electrode
3 Mica Sheet
4 Cubic Inner Anvil
5 Pressure Medium
6 Furnace
7 Sample
8 Thermocouple

Figure 3.10 Side view of one of the ultra pressure machines used in Japan. The enlargement shows a cross-section of the furnace assembly inside the inner tungsten carbide cubic anvils. From Ito *et al* (1984).

34

Disadvantages: Expensive to build and high operating costs. The pressure calibration is highly uncertain in some cases.

Typical applications: Phase transitions in the mid- to deep-mantle. Equations of state of minerals. Melting relations of mantle compositions to 300 kbar.

Description: There are many individual types of multi-anvil system. Their unifying characteristic is that pressure is generated by compression in four or more directions. The common types are tetrahedral (four directions) and cubic (six directions). A schematic drawing of one type is shown in Figure 3.10. Because of the mechanical complexity the reader is referred to the references listed in Liebermann *et al.* (1985) for a description of individual types.

Machine:

Shock tube

	P	*T*
Range:	Up to 5 Mbar	Up to 4,000°C
Precision:	±6%	?
Accuracy:	±6%	?

Sample volume: about 0.2 ml

Original references: Walsh & Christian (1955).

Working references: Ahrens (1980) is a good general description, Rigden *et al.* (1984) describe density measurements on preheated liquid samples.

Advantages: The best current method to generate very high *P* and *T* simultaneously.

Disadvantages: In this dynamic method the sample *P* and *T* change from ambient to maximum in microseconds, so there is always uncertainty about attainment of equilibrium. Samples usually do not retain their high pressure phase assemblage after the experiment, and details of the crystal structure cannot be determined.

Typical applications: The major application has been in measuring densities of minerals (and recently, liquids) and phase changes at the very high *P* and *T* conditions of the lower mantle and core of the Earth. The pressure–density information can be compared directly with seismic results to constrain mantle and core compositions. Recent experiments have been used to determine the relation between volatile loss and impact velocity of carbonaceous meteorites

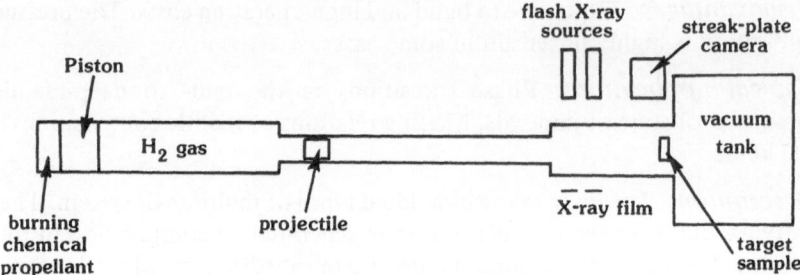

Figure 3.11 Light-gas gun for shock-wave experiments.

(Tyburczy & Ahrens 1986). Stolper *et al.* (1980) determined the relative densities of crystals and basaltic liquid at middle mantle pressures.

Description: The shock tube is used to fire a projectile at a sample target (Fig. 3.11). In the apparatus shown, the shock is generated in two steps. First a piston driven by exploding chemical propellant compresses hydrogen gas to high pressure. The gas in turn accelerates the projectile into vacuum. Pressure is generated by the shock wave transferred from the projectile to the sample. In the experiment shock velocity, U, and particle velocity, u, are measured. The shock velocity is determined by the time it takes the shock to go from the front to the back of the sample, and is measured by a streak plate camera and a system of mirrors on the sample. Particle velocity is proportional to the projectile velocity which is measured using nanosecond flash X-ray sources.

3.6 Machines for thermodynamic measurements

All of the techniques we have discussed up to now refer to experiments in which transformations from one phase to another are studied under variable P–T conditions. In addition to these direct experimental determinations, thermochemical measurements of enthalpy, entropy, and heat capacity provide us with the means to predict phase transformations without, in principle, observing them experimentally. These techniques are particularly useful if direct experiments are very difficult to perform, perhaps because of extreme P–T conditions or slow reaction rates. Some examples of the interrelationship between thermochemical and phase equilibrium measurements are given in Chapter 4.

3.6.1 Enthalpy determinations

The enthalpies of many minerals have been determined by solution calorimetry. In this method the heat evolved when the mineral is dissolved in a suitable solvent is measured precisely and compared with that given out when the constituent oxides of the mineral are dissolved. Original designs used hydrofluoric acid as the solvent in the temperature range 20–90°C (Torgeson & Sahama 1948, Robie & Hemingway 1972). Most recent data have been obtained at high temperatures in an oxide melt solvent.

Machine:

High temperature reaction calorimeter

	P	T
Range:	1 atm	600–1,000°C
Precision:	± 1%	in enthalpy of solution
	± 500 J	in heats of formation from oxides

Original references: Calvet & Prat (1954); Kleppa (1960).

Working references: Kleppa (1972), Kleppa (1976) for application to mineralogy. Akaogi *et al*. (1984) for application to very high pressure phases.

Advantages: Can operate with very small samples (20–50 mg), which is important for synthetic materials. Dissolution of refractory phases is usually achievable. Heat effects generally are a linear function of amount of solute. Works well over a wide temperature range. Precision comparable to hydrofluoric acid technique.

Disadvantages: Cannot readily work with any hydrous or carbonated mineral since these decompose before dissolution.

Typical application: Heat of dissolution of silicate in a solvent (generally $Pb_2B_2O_5$ melt) is compared with heats of constituent oxide dissolution to get heat of formation data.

Description: The apparatus consists of a block (Fig. 3.12) of nickel, nickel alloy or some other material stable at high temperature. It is surrounded by a thick layer of insulation and kept at constant temperature by two or three heating elements. Inside the block the sample chambers are each surrounded by a thermopile of 50 to 100 Pt–Rh thermocouples connected in series. Long silica glass tubes with an Au or Pt crucible inside are placed in each side of the calorimeter. One crucible contains about 30 g of $Pb_2B_2O_5$ with the sample suspended on a Pt cup attached to the manipulation tube. The other side acts as a reference. When thermal equilibrium is reached (after a few hours) the sample is gently stirred into the melt. This produces a heat effect which causes a slight change in the temperature of the sample side of the calorimeter. The temperature change is measured by the thermopile as a signal relative to the reference thermocouples. The signal is amplified, recorded, and integrated

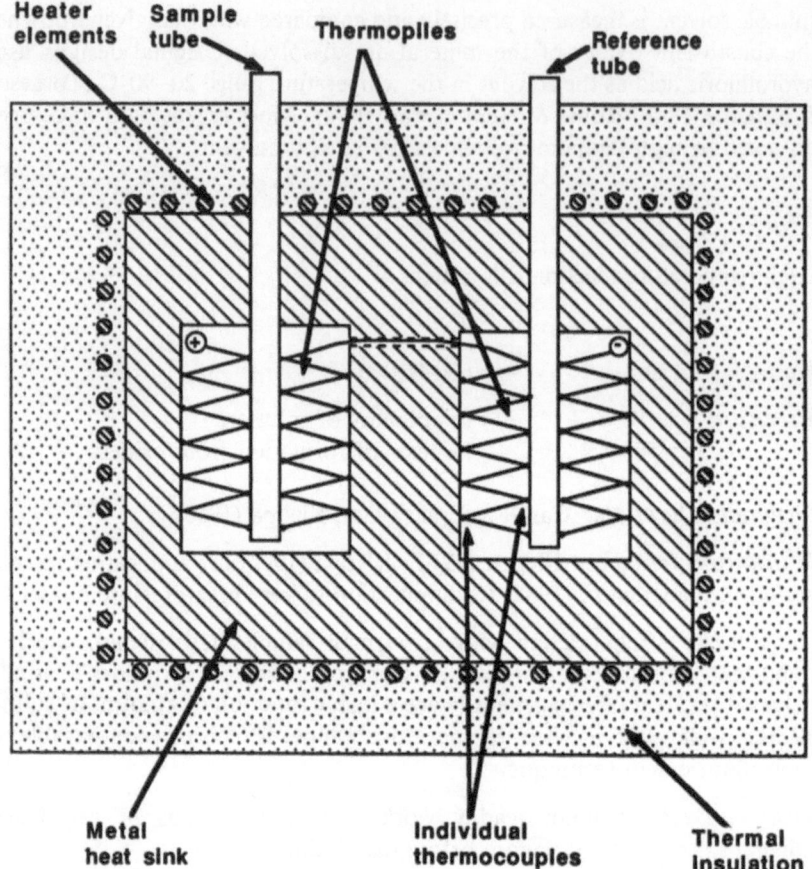

Figure 3.12 Schematic diagram of calvet-type twin macrocalorimeter for use at temperatures up to 800°C.

through the time taken to return to thermal equilibrium (about 45 min). The integrated signal is proportional to the heat of dissolution. The instrument is calibrated by dropping weighed pieces of Au or Pt from room temperature into the calorimeter and measuring the signal which corresponds to these known heat effects.

3.6.2 Heat capacity determination

Heat capacity measurements, at both low and high temperatures, comprise important constituents of the thermodynamic properties of minerals. The heat capacity is needed to extrapolate enthalpy up and down temperature:

$$H_{T_2} - H_{T_1} = \int_{T_1}^{T_2} C_p \, dT$$

38

and measurements at temperatures as close to 0 K as possible are needed to obtain the "third-law"entropies of minerals:

$$S_T = \int_0^T \frac{C_p}{T} \, dT$$

Drop calorimetry: The Calvet type calorimeter of Figure 3.12 may be used for drop calorimetry of samples from room temperature to calorimeter temperature. The heat effect constitutes the heat absorbed by the sample in heating up. Instruments specifically built as drop calorimeters employ a high temperature furnace with the sample being dropped to room temperature or $0°C$.

Machine:

Drop calorimeter

Temperature: 25–2,500°C

Precision in $H_{T_2} - H_{T_1}$: ±0.1%

Precision in heat capacity: ±1%

Uncertainty in heat capacity: ±1–2%

References: Hultgren *et al.* (1959); Douglas & King (1968); Robie (1987) recent summary.

Advantage: Operation to very high temperature.

Disadvantages: Large sample size. Several grams are required for good precision. Because there is a fairly large temperature difference between the drop temperature and the calorimeter temperature, this method gives an integrated heat effect ($H_{T_2} - H_{T_1}$) rather than the heat capacity directly. In order to obtain the latter, one has either to plot $H_{T_2} - H_{T_1}$ against T_2 and graphically differentiate, or fit a curve to ($H_{T_2} - H_{T_1}$) as a function of temperature and differentiate the fitted equation. This results in a larger uncertainty in C_p than in ($H_{T_2} - H_{T_1}$) ± 1% as compared to ±0.1%.

Description: Furnace is essentially of standard 1 atm quench type. Calorimeter cell can be of several different designs. In one of the most widely used types (Douglas & King 1968) the sample is dropped from high temperature into a metal well which is surrounded by ice. The heat given out by the sample melts a certain volume of ice. The volume of water produced (and ice decomposed) is then carefully measured by displacement of mercury. An alternative method (Hultgren *et al* 1959) uses the heat given up by the sample to cause expansion of diphenyl ether. This liquid has a very high coefficient of thermal expansion at the temperature of the calorimeter (27°C), and precise measurement of its volume before and after the experiment is a measure of the heat evolved as the sample cools.

3.6.3 Differential scanning calorimetry

The differential scanning calorimeter (DSC) is a rapid method of direct heat capacity measurement to moderate temperatures.

Machine:

Differential scanning calorimeter

Temperature: − 100 to + 750°C

Precision in heat capacity: ± 1%

Uncertainty in heat capacity: Approximately ± 1%

Original reference: Watson *et al.* (1964)

Working references: Commercially available from Perkin−Elmer Co. See Krupka *et al.* (1979) for application to minerals.

Advantages: Small samples (20 mg) are required, so the method is readily applicable to synthetic materials. Direct rapid measurement of heat capacity with precision at least as good as drop calorimetry.

Disadvantage: Temperature limit of about 1,000 K.

Operation: The DSC operates with a twin cell, one of which is a reference in the same kind of way as the high temperature reaction calorimeter. The calorimeter measures the difference in energy required to keep the sample at some temperature T and that required to keep the reference at the same temperature. As temperatures of both sides are raised (at about 10°C/min) signals representing the sample and reference temperatures are fed to a differential temperature amplifier. The differential temperature amplifier output then computes the power to the electrical heaters in the sample side and reference side. This adjustment corrects the temperature difference between sample and reference. The difference in heat input between sample side and reference side over a small temperature increment gives the heat capacity of the sample.

3.6.4 Low temperature calorimetry

Low temperature heat capacity measurements are particularly important for the derivation of "Third Law" entropies. Although many different designs have been used since the 1920s, the adiabatic calorimeter is generally employed in mineralogic studies.

Machine:

Adiabatic calorimeter

Temperature: 10 to 380 K

Uncertainty in heat capacity: ±0.15% between 25 and 380 K
±1% at 15 K

References: Westrum *et al.* (1968a), Robie & Hemingway (1972)

Working references: Robie & Hemingway (1972) give a concise but detailed description of the operation of the calorimeter. See, for example, Robie *et al.* (1978) for mineralogic applications; Robie (1987) for detailed description.

Advantages: Very precise measurements at low temperatures. Direct heat capacity determinations.

Disadvantages: Large sample size (several grams) is required. Although adiabatic calorimetry can operate at temperatures higher than 100°C, experimental difficulties increase with increasing temperature.

Description: The calorimeter itself is a small copper piece inside the complex looking apparatus of Figure 3.13. It contains the sample, a platinum resistance thermometer, and an electrical heater. The calorimeter is suspended on a nylon line inside the adiabatic shield. The latter acts to thermally isolate the calorimeter from its surroundings. The submarine, consisting of calorimeter, adiabatic shield, isothermal shield, and liquid helium reservoir (Fig. 3.13) is then lowered into the Dewar flask. The Dewar is filled with liquid nitrogen which, with the aid of a vacuum pump, cools the submarine to 52 K. The helium reservoir is filled with liquid helium (at 4.2 K), and the adiabatic shield and calorimeter are winched up on the nylon line until they are in contact with the reservoir. When temperature in the calorimeter has dropped to 10 K the shield and calorimeter are winched down to thermally isolate them from the reservoir. Measurements are made by determining how much energy from the electrical heaters is required to raise the temperature of the calorimeter a certain number of degrees, generally of the order of one degree at 10 K, three degrees at 30 K and five degrees at temperatures above 50 K. The temperature of the adiabatic shield is controlled with its own heater to be the same as that the calorimeter. Thus, heat flow out of the calorimeter is negligible.

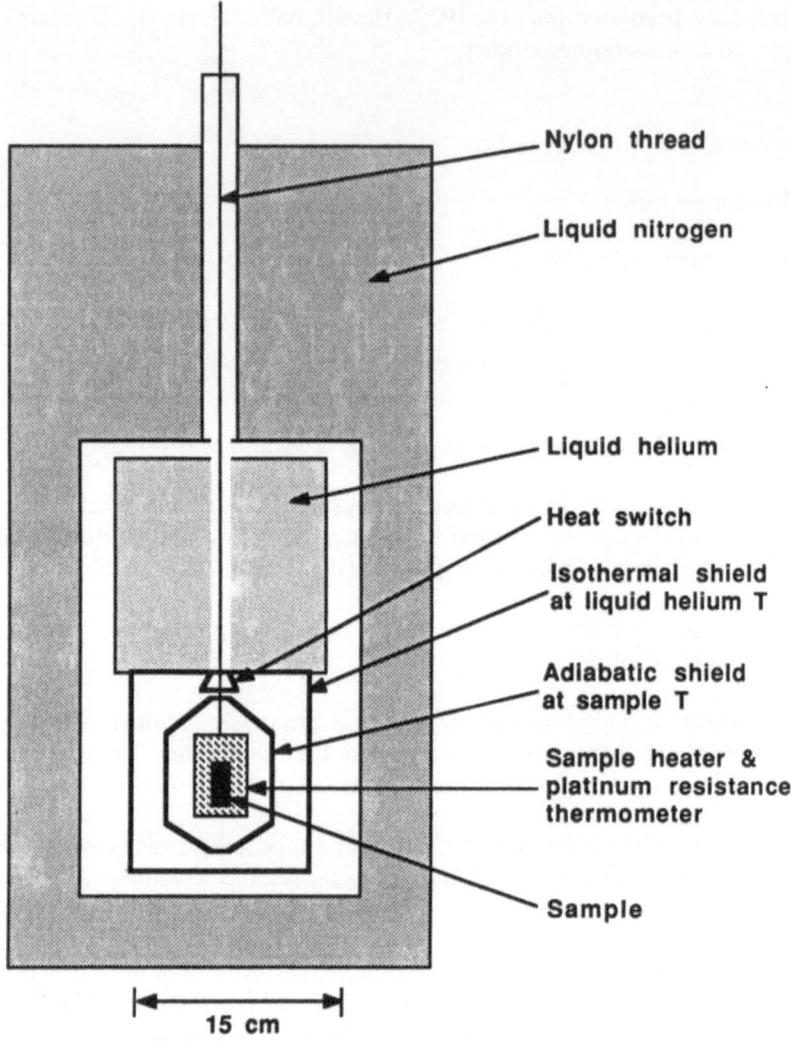

Figure 3.13 Schematic diagram of a helium cryostat for heat capacity measurements between 10 and 380 K.

3.6.5 *Activity measurements*

The calorimetric methods discussed in the previous section provide means of obtaining enthalpy and entropy data. Of equal interest are methods of determining relative free energies of phases or partial molar free energies of components under known *P*, *T* and composition conditions. Although several different experimental methods of doing this will be discussed in Chapters 4 and 5, it is appropriate here to discuss direct, *in situ* measurements with solid electrolytes.

Suppose we have coexistence, at equilibrium, of pure hematite and pure magnetite:

$$6Fe_2O_3 \rightleftharpoons 4Fe_3O_4 + O_2$$
$$\text{hematite} \qquad \text{magnetite}$$

The condition of equilibrium, at some P and T, with both solids pure and with a 1 bar and T standard state for oxygen is:

$$\Delta G_{P,T}^0 = -RT \ln a_{O_2}$$

Direct measurement of oxygen activity (a_{O_2}) would therefore give the standard state free energy change of the reaction.

Machine:

Oxygen activity sensor

Pressure: 1 atm

Temperature: 700–1,200°C

Precision in measuring oxygen activity: ±0.05 log units

Accuracy in oxygen activity: ±0.1 log units

Original reference: Kuikkola & Wagner (1957)

Working references: Sato (1971). See Elliott *et al.* (1982) and Arculus & Delano (1981) for applications to intrinsic oxygen fugacity determination of rocks. See Huebner (1987) for detailed description.

Advantages: Precise determination of free energy data for redox equilibria. "Intrinsic" f_{O_2} determinations for rocks and minerals. Small sample size. Rapid equilibration times (typically a few hours).

Disadvantages: Cells can be poisoned by contaminants such as iron. Oxygen leakage through the wall can cause drift. At low f_{O_2}'s electronic conduction short-circuits the important oxygen ion conduction mechanism.

Applications: Determination of equilibrium oxygen activity (fugacity) for solid buffers. Measurement of intrinsic oxygen fugacities of minerals and rocks.

Description: A typical cell (Fig. 3.14) consists of a plug of ZrO_2 which is doped with Y_2O_3 or CaO so that it conducts oxygen ions (O^{2-}). The plug is

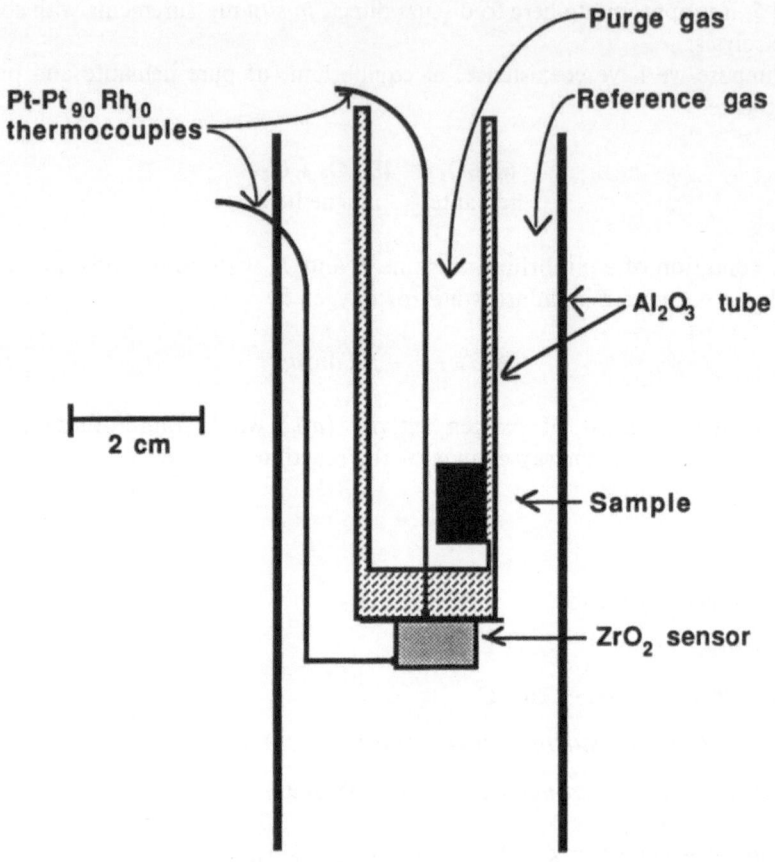

Figure 3.14 Sketch of an oxygen activity sensor for use at atmospheric pressure. The Al_2O_3 tube has a plug of ZrO_2 electrolyte at one end and the inside of the cell is isolated from the reference gas stream.

sealed into an alumina tube so that there is no exchange of gas between inside and outside and coated with platinum for good electrical contact. The sample, in an unwelded capsule or pressed pellet, is placed inside the cell and kept from direct contact with the electrolyte. The inside of the cell is then flushed with a high purity inert gas such as N_2 or Ar, and when all air is removed the inside is sealed. The cell is then placed into a 1 atm gas mix furnace and a reference gas (CO–CO_2 mixture) passed across the outside. The difference in oxygen activity between inside and outside creates an emf difference which can be measured with a high impedance millivoltmeter. The measurement is made across the leads of the inside and outside thermocouples which are in contact with the electrolyte. Reversal can be obtained in two ways. One way is to pass a reducing gas into the sample chamber, seal off, and obtain a stable emf. Repeating with an oxidizing gas gives the approach to equilibrium from both sides. Alternatively, attaching a battery across the inner and outer thermocouple leads enables small amounts of oxygen to be conducted into or out of the cell. Approach from both directions gives a reversal.

When the position of equilibrium is obtained the oxygen activity (fugacity) of the sample is obtained from the measured emf using the Nernst equation:

$$E = \frac{RT}{4F} \ln \left(\frac{a_{O_2}^{\text{sample}}}{a_{O_2}^{\text{reference}}} \right) \text{ volts}$$

F is the Faraday constant and the factor 4 takes account of the fact that each oxygen molecule produces $2O^{2-}$ ions so that four electrons are transported across the cell.

The experimental arrangement of Figure 3.14 works well if the sample has a moderate oxygen buffering capacity, but more sophisticated arrangements are required for intrinsic f_{O_2} measurements. These involve the use of a reference gas very close to the sample f_{O_2} to minimize oxygen leakage and measurement relative to a reference cell (double-cell method) to minimize drift (Sato 1971, Arculus & Delano 1981).

Other solid electrolytes which have potential applications in geology are sulfur and fluorine conductors (Sato 1971) and the high temperature pH electrode (Niedrach 1980). Since the latter is already in limited use, we will briefly discuss it.

Machine:

High temperature pH electrode

Pressure: liquid–vapor curve for H_2O

Temperature: 25–300°C

Precision: ±5% in pH and improving

Accuracy: ±5% in pH and improving

Original reference: Niedrach (1980).

Working references: Danielson *et al.* (1985); see Bourcier *et al.* (1987) for detailed description.

Advantages: Provides direct pH measurement at elevated temperature in hydrothermal solutions.

Disadvantage: No precise calibration at high temperatures. Non-reproducible behavior unless sensor is carefully pretreated at 250–300°C

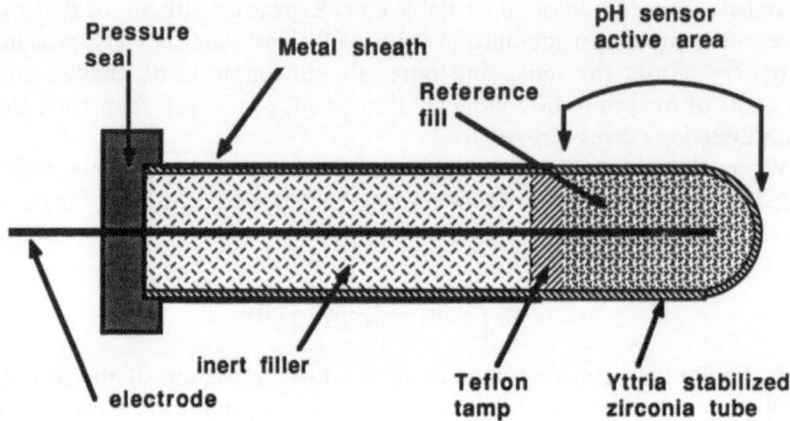

Figure 3.15 Cross-sectional view of the pH electrode for use at pressures of a few hundred bar and temperatures to 300°C.

Description: Makes use of the same Y_2O_3-stabilized ZrO_2 ceramic as the oxygen activity sensor (Fig. 3.15). Since this ceramic is an oxygen ion conductor it provides the possibility to measure pH indirectly through the equilibrium:

$$2H^+ \;+\; O^{2-} \;\rightleftharpoons\; H_2O$$

$$\text{solution} \quad \text{solution} \quad \text{solution}$$

The inside of the sensor contains either an electrolyte solution or a mixture of Cu and Cu_2O with a copper wire electrode in it (Fig. 3.15). This will operate at temperatures up to 300°C in a pressure vessel for periods of days to weeks. The reference electrode is an external Ag/AgCl electrode which runs at pressure-vessel pressure but at 25°C temperature (Danielson *et al.* 1985).

Recent advances in this technology which are also of geologic interest are the development of Eh sensors for high temperature systems (Macdonald *et al.* 1980, 1981) and reported improvements in precision of the high temperature pH sensor (Hettiarachi & Macdonald 1984).

3.7 Conclusions

In this chapter we have summarized a wide spectrum of techniques which are available to the experimental geochemist. Although the most used ones are described, many others are sufficiently limited in use that we did not feel it necessary to describe them in detail. We will, however, refer to a number of these other methods as they arise in the succeeding chapters in, we hope, enough detail for the reader to get an idea of how they work.

4 Metamorphic experiments on solid–solid reactions

4.1 Introduction

Many metamorphic rocks crystallize or recrystallize under conditions which generate an essentially volatile-free mineral assemblage. As examples, one might consider rocks of the granulite facies which typically produce orthopyroxene–clinopyroxene–plagioclase $\pm$ garnet $\pm$ quartz assemblages in mafic compositions and quartz–plagioclase–alkali feldspar–sillimanite –garnet $\pm$ cordierite assemblages in pelitic compositions. Of even wider distribution than granulites are the recrystallized rocks of the upper mantle which are found as xenoliths in mantle-derived volcanics. Peridotites and eclogites are the commonest of such volatile-free xenoliths.

These anhydrous rocks did not necessarily crystallize under fluid-absent conditions, of course; the absence of carbonates or hydrates simply requires that P, T a_{H_2O} and a_{CO_2} were inappropriate for the stability of volatile-containing minerals. The observation means, however, that one does not have to take explicit account of the activities of H_2O and CO_2 in experiments on these rocks.

A number of important questions has arisen concerning the origins of high-temperature metamorphic assemblages. The almost exclusive occurrence of granulites in Precambrian metamorphic belts has led to the suggestion that Precambrian metamorphism may have involved a higher geothermal gradient than Phanerozoic metamorphism. The spinel lherzolite–garnet lherzolite transition in ultramafic rocks has attracted attention because it affects upper mantle density (slightly) and the trace element contents of mantle-derived melts (significantly). There has been considerable discussion as to whether or not eclogites are stable under lower crustal conditions, conditions which yield almost no reaction on experimental time scales. A reliable way of extrapolating experimental results to these conditions would provide an answer. Garnets in high grade metamorphic rocks are commonly zoned with respect to Ca, Fe, and Mg. The zoning patterns contain P–T-time information for the mineral assemblages, but getting at the information is impossible without well calibrated geothermometers and geobarometers. These are just a few questions; the reader can probably come up with many more.

The integration of phase equilibrium and thermodynamic data is important for the study of subsolidus reactions. Adiabatic and drop calorimetry yield very precise measurements of heat capacities and third law entropies (S) of pure minerals, and X-ray crystallography provides very accurate volume measurements. Nevertheless, solution calorimetry gives relatively imprecise measurements of the enthalpies (H) of minerals, and any residual disordering contribution to entropy cannot be detected calorimetrically. Therefore, if we consider a solid–solid reaction such as:

$$NaAlSi_3O_8 \rightleftharpoons NaAlSi_2O_6 + SiO_2 \qquad (4.1)$$

$$\text{albite} \qquad \text{jadeite} \qquad \text{quartz}$$

calorimetric and crystallographic data give us:

$$\Delta G^0_{P,T} = (G^0 \text{ products} - G^0 \text{ reactants}) = \Delta H^0_{P,T} - T \, \Delta S^0_{P,T} \text{ (imprecise)}$$

$$\left(\frac{\partial \, \Delta G^0}{\partial T} \right)_P = - \Delta S^0 \text{ (precise if there is no residual disorder)}$$

$$\left(\frac{\partial \, \Delta G^0}{\partial P} \right)_T = \Delta V^0 \text{ (precise)}$$

A phase equilibrium study of Reaction (4.1), while not yielding any direct thermodynamic data, gives the relative free energies ($\Delta G_{P,T}$) of products and reactants with much higher precision than calorimetry. The two approaches, calorimetric and phase equilibrium, are obviously complementary.

In addition to generating basic information on relative stabilities of pure minerals, phase equilibrium experiments can also be used to measure the activities of components in multicomponent phases. Experiments of the latter type, which will be discussed in detail below, provide the building blocks of models of complex multicomponent phases.

Good experiments on solid–solid reactions have all the ingredients discussed in Chapter 2. Well characterized reactant and product materials are mixed, run under precisely controlled conditions, quenched, and examined. If the phases are pure, the direction of reaction is determined by growth and dissolution of stable and unstable assemblages, respectively. Further characterization of products might require determination of ordering states in minerals such as albite which exhibit variable Al–Si disorder. Experiments on solid solutions generally require the determination of mineral compositions as well as ordering states after the run.

Under conditions close to the melting point experiments on solid–solid reactions can be performed in sealed capsules with fluid completely absent. At temperatures of more than about 100°C below melting, however, reaction rates in the absence of fluid become extremely low. The way around low reaction rates is to add a flux such as H_2O or an H_2O-CO_2 mixture (e.g., Johannes *et al.* 1971, Boettcher 1970, Jenkins & Newton 1979). An H_2O-rich flux dramatically speeds up reactions by providing a transporting medium in which reacting minerals dissolve to some extent. This approach is good at temperatures well below melting but is hampered by the fact that H_2O depresses the melting point substantially, so that there is a well defined upper temperature limit to the method. In addition, most systems produce hydrous minerals at low temperature, so that there is a low temperature limit to the method as well. The result is that many solid–solid reactions have a "window" at moderate temperatures in which an H_2O-rich flux works and a narrow high temperature interval in which the reaction will take place in the absence of fluid. In order to bridge the gap between these two regions, another flux must be used. A flux consisting of lead silicate melt has worked successfully in a number of systems (e.g., Gasparik 1984a, Gasparik & Newton 1984). Lead is insoluble in most common minerals (except feldspar), so it does not affect the $P-T$ positions of many reactions significantly. Furthermore, the lead silicate melt dissolves Ca, Mg, Al, etc., in sufficient amounts to promote reaction rates.

4.2 Reactions involving pure phases

Jadeitic pyroxene is an important consitituent of rocks crystallized at high pressures and low temperatures such as those of the Franciscan of California. Reaction 4.1:

$$\text{Albite} \rightleftharpoons \text{Jadeite} + \text{Quartz}$$

is a simple analog for the conditions under which albitic plagioclase breaks down to produce clinopyroxene and quartz. Holland (1980) studied this reaction over the temperature range 600–1,200°C in a piston-cylinder apparatus using an NaCl pressure medium (Section 3.4). His starting materials were mixtures of synthetic high albite, natural jadeite, and natural quartz. Water was used as a flux at 600°C, while at 800°C and above the charge was just moistened slightly before sealing in platinum capsules. Under the latter conditions, the small amount of water was sufficient to produce a few per cent of hydrous silicate melt which acted as a flux for the reaction. Holland's results (Fig. 4.1) demonstrate that the reaction can be tightly reversed over the 600–1,200°C temperature range with relative uncertainties of between ±250 and ±100 bar.

High temperature solution calorimetic measurements of Charlu *et al.*

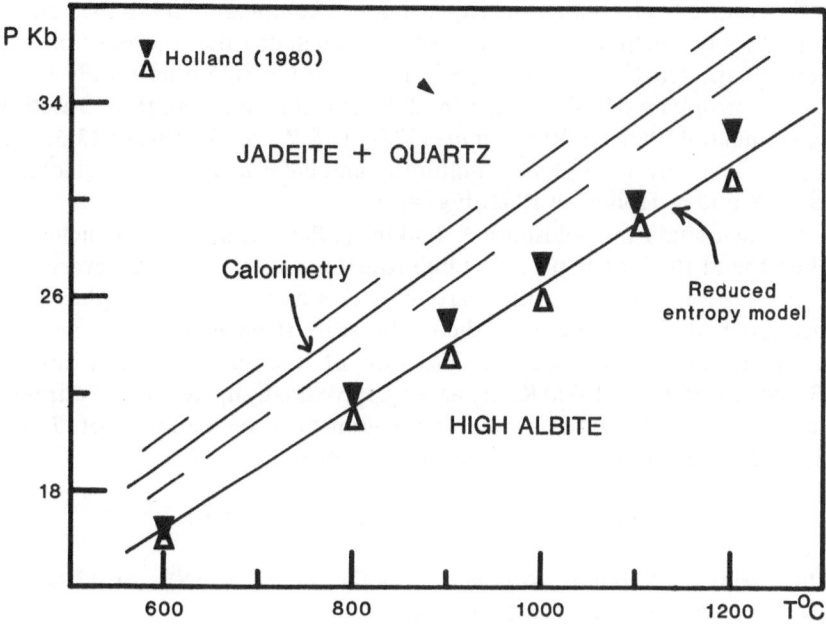

Figure 4.1 Stability field of high albite. Filled and open triangles refer respectively to Jd + Q stable and albite stable. Best-fit curve requires entropy of Al–Si disorder in high albite of about 13.6 J/K (see text). From Holland (1980).

(1975), Newton *et al.* (1980), and Wood *et al.* (1980) give ΔH^0 of reaction at 1 bar and 1,000 K of $-8,910 \pm 960$ J if high albite is the reactant. The entropy of reaction, based on calorimetry, would be -29.7 J/K if Al–Si disorder in albite were neglected. In practice, Al and Si are known to be disordered in high albite, and addition of the entropy contribution corresponding to complete disorder lowers ΔS^0 at 1 bar and 1,000 K to -48.5 J/K (Robie *et al.* 1978). The volume change of reaction is -17.34 ml at 25°C and 1 atm pressure and ΔC_p of reaction is almost zero at 1,000 K. These data can be used to calculate the approximate position and slope of the boundary from:

$$\Delta G^0_{P,T} = \Delta H^0_{1,T} - T\,\Delta S^0_{1,T} + P\,\Delta V^0_{1,298} = 0$$

$$P = [48.5\,T\,(K) - 8190]/1.734 \text{ bar}$$

As can be seen from Figure 4.1, the calculated curve overestimates the stability field of high albite by several kbar. The combined uncertainties in enthalpy and entropy correspond to about 2 kJ or a little over 1 kbar in pressure. The discrepancy between experimental and calorimetric data cannot, therefore, be solely explained by uncertainties in the latter. Adding in the contributions of thermal expansion and compressibility to the volume term does not help much either. Holland concluded that the position and slope of the reaction could be consistent with the calorimetry only if the actual entropy of Al–Si disorder in albite is rather less than the theoretical maximum (tabulated in Robie *et al.* 1978) of 18.7 J/K. A value of 13.56 J/K is consistent with both phase equilibrium and calorimetric results (Holland 1980; Wood & Holloway 1982: Fig. 4.1).

A recent study by Goldsmith & Jenkins (1985) confirms this conclusion. They found that the high–low albite transformation could be reversed at high pressures using a sodium carbonate flux and that it takes place with increasing Al–Si disorder over the temperature range 680–800°C. Newton *et al.* (1980) found that the enthalpy of the low-high transition is 13,400 ± 1,250 J at 1,000 K. If, as an approximation, we treat the transition as a discontinuous one at an equilibrium temperature of 740°C (1,013 K) the following entropy change is obtained:

$$\Delta S^0_{disorder} = \Delta H_{dis}/T_{dis} = 13.23 \pm 1.24 \text{ J/K}$$

This result corresponds very well with the value of ΔS^0_{dis} of 13.56 J/K derived from the albite breakdown Reaction 4.1.

Another example of the interrelationship between phase equilibrium and calorimetric data is provided by Goldsmith's (1980) study of the reaction:

$$3CaAl_2Si_2O_8 \rightleftharpoons Ca_3Al_2Si_3O_{12} + 2Al_2SiO_5 + SiO_2 \qquad (4.2)$$

anorthite $\qquad$ garnet $\qquad$ kyanite $\quad$ quartz

The experiments were performed in a piston-cylinder apparatus with oxalic acid being used to produce an H_2O–CO_2 flux at temperatures of 1,350°C and below.

Calorimetric data yield enthalpy and entropy of reaction at 1,000 K and 1 atm of $-40,600 \pm 6,700$ J and -122.6 J/K, respectively. The volume change at 25°C and 1 atm is -66.3 ml or -6.63 J/bar. These data are readily turned into an equation for the univariant reaction:

$$P = [122.6\ T(K) - 40,600]/6.63 \pm 1 \text{ kbar calorimetric uncertainty} \qquad (4.3)$$

As can be seen from Figure 4.2, the calculated curve is well below the observed boundary and the uncertainties in the calorimetric data do not cover the apparent error. The situation can be improved to some extent by taking account of thermal expansions and compressibilities of reactant and product phases. To a good approximation the volume of a mineral in the P–T regime of interest is given by:

$$V_{P,T} = V_{1,298}[1 + \alpha(T - 298) - \beta P]$$

52

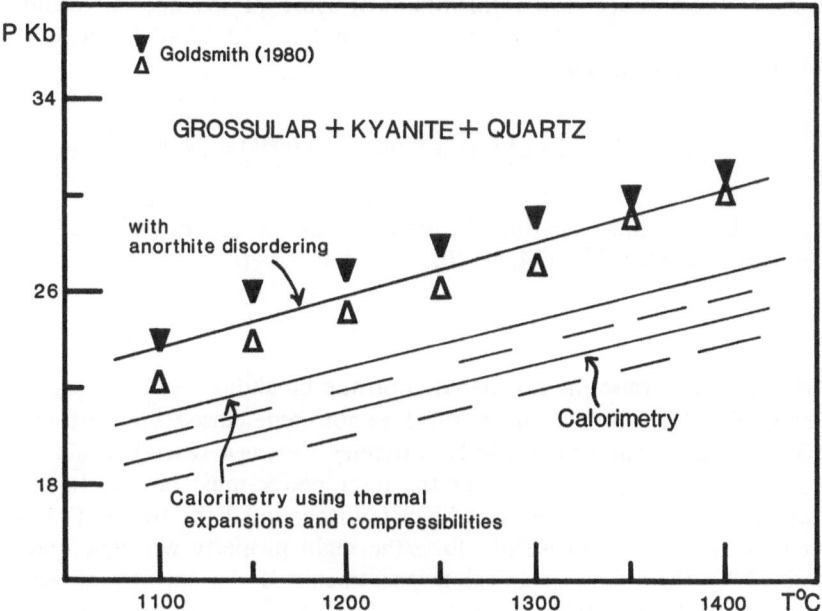

Figure 4.2 Calculated position of the anorthite breakdown reaction. Lowermost solid line is calculated from calorimetric data with dashed lines indicating uncertainty limits. Addition of the effects of compressibility, thermal expansion and Al–Si disorder raise the calculated curve into agreement with phase equilibrium experiments. Filled triangles = anorthite unstable; unfilled triangles = anorthite stable.

where $V_{1,298}$ is the volume at 25°C and 1 bar and α and β are thermal expansion and compressibility, respectively. The contribution of volume to the free energy of a phase at pressure P relative to free energy at 1 bar is calculated from:

$$G_{P,T} - G_{1,T} = \int_1^P V \, dP = P[1 + \alpha(T - 298) - \beta P/2] V_{1,298}$$

This equation is the next higher approximation to the one embodied in Equation 4.3, in which volume is assumed to be independent of P and T. The volume of 3 mol of anorthite at 298 K and 1 bar is 302.4 ml (30.24 J/bar) and it has α of $1.5 \times 10^{-5} \deg^{-1}$ and β of $1.1 \times 10^{-6} \text{bar}^{-1}$ (Wood & Holloway 1984). At an average T and P of 1,500 K and 25 kbar, respectively, this gives:

$$\int_1^P V \, dP = 30.24P(1 + 0.018 - 0.014) = 30.36P$$

At 25°C, the products have total volume of 236.1 ml, average expansibility of $2.7 \times 10^{-5} \deg^{-1}$ and compressibility of $0.6 \times 10^{-6} \, bar^{-1}$. Hence we obtain, for the products:

$$\int_{1}^{P} V \, dP = 23.61 P \, (1 + 0.0325 - 0.0075) = 24.20 P$$

As a better approximation the volume change in Equation 4.3 may therefore be replaced by $(24.2 - 30.36)$, -6.16 J/bar:

$$P = [122.6 \; T(\text{K}) - 40{,}600]/6.16 \pm 1 \text{ kbar} \tag{4.4}$$

The corrections raise the calculated anorthite breakdown pressure by 1 to 2 kbar (Fig. 4.2), but the curve still does not come close to that observed experimentally. The heat capacity difference between reactants and products is extremely small, so that the discrepancy must be due to some property of one of the phases which is unaccounted for by the thermochemical data. In the case of albite the main property which is unconstrained by the thermochemical measurements is the entropy of Al–Si disorder in albite. Residual, quenched in, disorder cannot be measured thermochemically, so it has to be estimated from an order–disorder model. The ideal random mixing model was shown, on the basis of phase equilibrium data, to overestimate the disordering contribution to high albite entropy. The opposite explanation probably holds in the case of anorthite. The entropy of anorthite used in the calculation (Robie *et al.* 1978) is based on the assumption of no Al–Si disorder since this phase is generally assumed to exhibit the "aluminum avoidance" principle. It has been shown experimentally, however, that anorthites quenched from high temperature do indeed have a few per cent of Al–Si disorder. The amount required to explain the phase equilibrium data is 1 to 2%, sufficient to add 1.8 to 4.0 J/K to the entropy of anorthite in the experimental temperature range. Thus, Equation 4.4 can be modified to take account of a small entropy of Al–Si disorder (10 J/K for 3 mol) and an allowance made for uncertainties in the enthalpy data to give:

$$P = [132.6 \; T(K) - 35550]/6.16 \text{ bar} \tag{4.5}$$

It can be seen from Figure 4.2 that the corrections provide a good fit to Goldsmith's experimental results.

This section has been devoted to two solid–solid reactions for which very good calorimetric and phase equilibrium data are available. Attempts to integrate the data into a coherent whole yield an appreciation of the thermodynamic properties of the minerals which goes beyond either calorimetric or phase equilibrium data on their own. Clearly both approaches must be considered in tandem if one is to understand how minerals behave at high pressures and temperatures.

4.3 Displaced equilibria and activity measurements

Part of the point of this chapter is to show that there are good ways of building up from a data base on pure phases to the calculation of phase relationships in multicomponent rocks. This process requires the determination of activities of major components in the solid solutions (albite–anorthite plagioclase, etc.) which occur in nature. One simple way in which solid–solid reactions are used to determine activities in multicomponent minerals is by the method of "displaced equilibrium." The approach involves first determining a univariant reaction involving the pure phase of interest, e.g.,

Albite ⇌ Jadeite + Quartz

An additional component, which is only soluble in the mineral of interest, is then added to the system. For example, addition of $CaMgSi_2O_6$ component to albite–jadeite–quartz assemblages produces a jadeite–diopside–clinopyroxene solid solution coexisting with essentially pure albite and quartz. The equilibrium composition of the clinopyroxene in the three-phase assemblage is then determined under $P–T$ conditions displaced from those of the pure end-member reaction. Determination of equilibrium composition at known P and T results in a direct measurement of the relationship between activity and composition from the relationship:

$$RT \ln a^{cpx}_{NaAlSi_2O_6} = \int_P^{P^0} \Delta V^0 \, dP = \Delta V^0_{P,T}(P^0 - P) \qquad (4.6)$$

In Equation 4.6, P refers to the pressure of the experiment at temperature T and P^0 to the equilibrium pressure for the end-member reaction at the same temperature. This type of experiment is particularly relevant to nature because few metamorphic pyroxenes approach pure jadeite in composition. High temperature granulite pyroxenes typically contain 4 or 5 mol% $NaAlSi_2O_6$, very low temperature Franciscan-type pyroxenes 70 to 80 mol% $NaAlSi_2O_6$, and eclogite pyroxenes about 50 mol% $NaAlSi_2O_6$. In the latter case, at temperatures less than $800°C$, the 50 mol% $NaAlSi_2O_6$, 50 mol% $Ca(MgFe)Si_2O_6$ composition is an ordered omphacite with $P2/n$ structure rather than a disordered high-temperature clinopyroxene.

Holland (1983) has determined a pressure–composition diagram for $NaAlSi_2O_6–CaMgSi_2O_6$ clinopyroxene coexisting with jadeite and quartz at $600°C$. His results are reproduced in Figure 4.3 and converted into a plot of

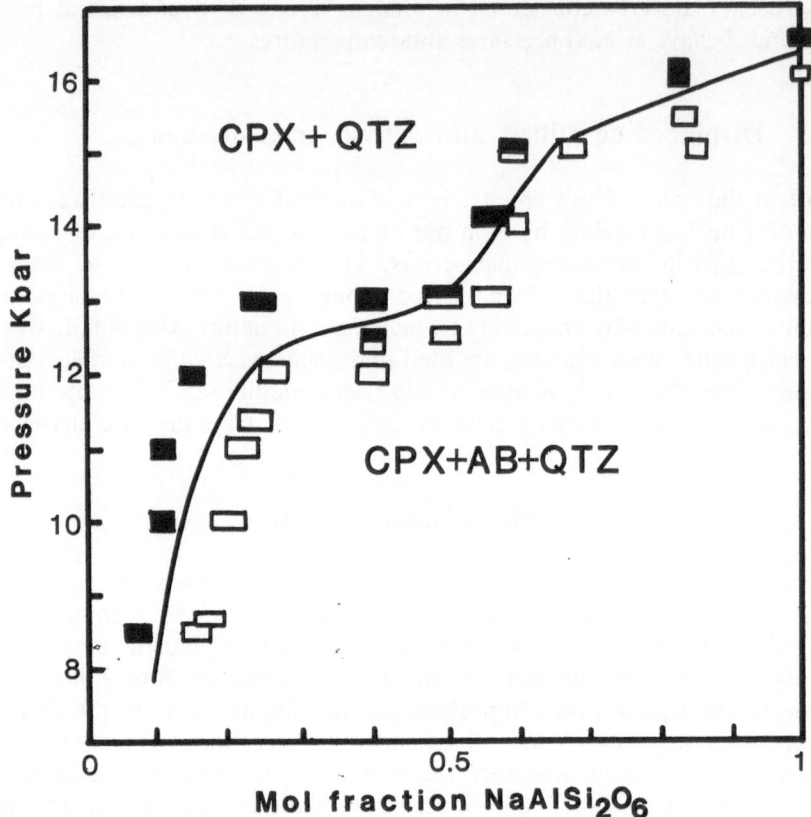

Figure 4.3 Experimental results displayed as a pressure–composition section at $600°C$. Open boxes and filled boxes are compositions of pyroxenes partially equilibrated from $NaAlSi_2O_6$-oversaturated and -undersaturated starting materials respectively. Half-filled boxes indicate no change after the run. Results from Holland (1980) and Holland (1983).

$(NaAlSi_2O_6$ activity$)^{1/2}$ versus composition using Equation 4.6, in Figure 4.4. The reason for taking the square root of activity is that with two cation sites involved in the mixing process of NaAl and CaMg, $a^{1/2}_{NaAlSi_2O_6}$ should be proportional to the mole fraction of $NaAlSi_2O_6$ if the solid solution is ideal (see below).

The starting clinopyroxenes for all of the experiments were synthesized at high temperatures and had the disordered $C2/c$ structure. They all retained this structure during the experiments despite the fact that the ordered omphacite structure is actually stable for compositions close to 50:50 at $600°C$. It can be seen, however, that the latter compositional region is one for which a pronounced kink occurs in the $a^{1/2} - X$ curve of Figure 4.4. Holland concluded that, although not in the omphacite structure, composi-

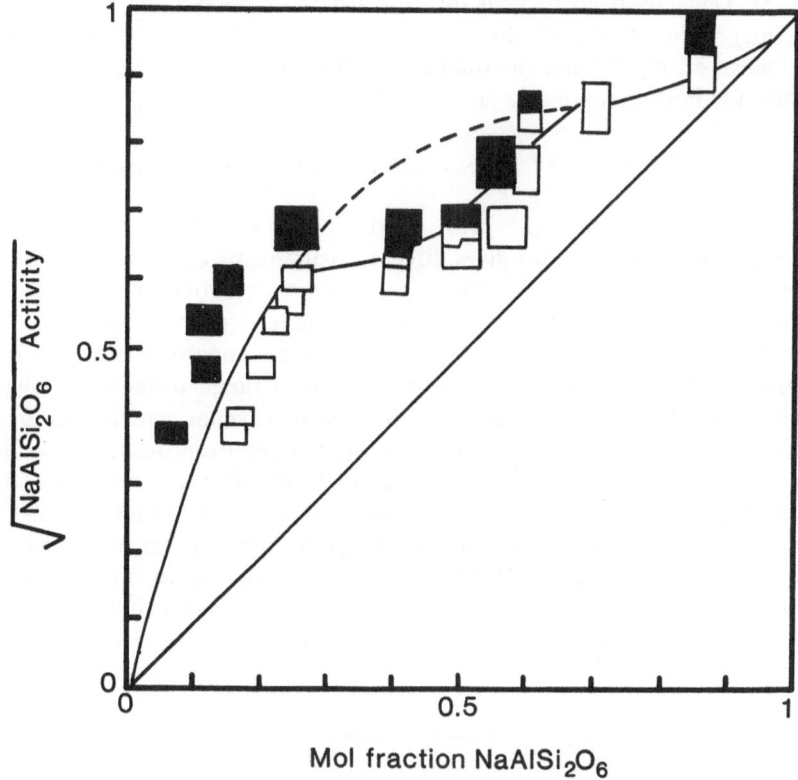

Figure 4.4 Activity–composition diagram derived from the data of Figure 4.3. The full line corresponds to the curve in Figure 4.3 and the dashed portion represents an extrapolation of the best-fit model to the disordered region. Diagonal line corresponds to ideal mixing.

tions close to 50 : 50 had undergone substantial short range ordering during the experiments and that the energetic effect of this was comparable to that associated with the long range $C2/c \rightarrow P2/n$ transformation.

The activity measurement may be used to obtain activity–composition relationships for the other component ($CaMgSi_2O_6$) and total excess free energy of the solid solutions in the following manner. The activity coefficient (γ) of the $NaAlSi_2O_6$ component is defined by

$$a^{cpx}_{NaAlSi_2O_6} = X^2_{NaAlSi_2O_6} \, \gamma_{NaAlSi2O6} \qquad (4.7)$$

The measured values of X and the calculated value of activity yield γ directly. The squared term arises from the fact that there are two sites of mixing in $NaAlSi_2O_6$–$CaMgSi_2O_6$ clinopyroxenes (Wood & Fraser 1976,

Ch. 3). The activity coefficients for the $CaMgSi_2O_6$ component are obtained by integration of the Gibbs–Duhem equation (Wood 1977 gives mineralogical examples) and the total excess free energy from the addition of values for both components as follows:

$$G^{xs} = X_{NaAlSi_2O_6} \, RT \ln \gamma_{NaAlSi_2O_6} + X_{CaMgSi_2O_6} \, RT \ln \gamma_{CaMgSi_2O_6} \quad (4.8)$$

Calorimetry fits into the picture, too. Solution calorimetry on diopside–jadeite solid solutions gives direct measurements of the excess enthalpies of the solid solutions. These measurements should correspond to experimental excess free energies provided there is no abnormal entropy effect associated with mixing. Figure 4.5 shows a comparison of excess free energies obtained at $600°C$ (ignoring the short-range order effect) with excess enthalpies determined on $C2/c$ clinopyroxenes which had been synthesized at $1,300°C$. It can be seen that, if the ordering phenomenon discussed above is factored out, then the forms of excess enthalpy and excess free energy terms are very similar. The latter, determined on $600°C$ pyroxenes, are however slightly smaller than the former obtained on pyroxenes synthesized at $1,300°C$. Some of the difference is due to a small amount of short-range order present in all $C2/c$ clinopyroxenes on this join. The short-range order interpretation put forward by Wood *et al.* (1980)

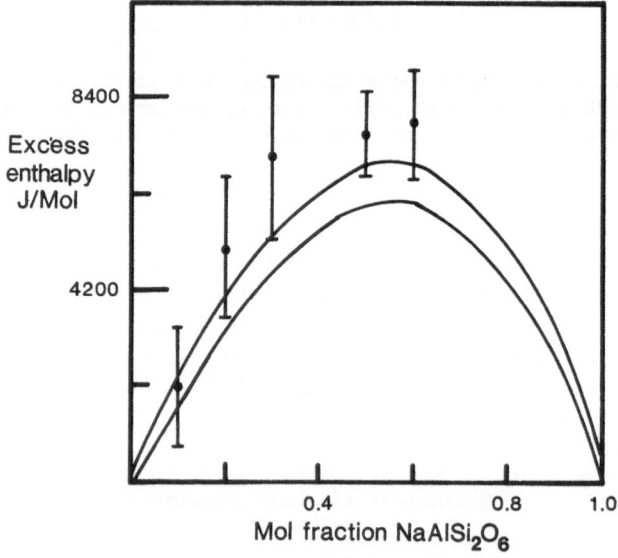

Figure 4.5 Comparison of measured excess enthalpies and excess free energies of NaAl-Si$_2$O$_6$–CaMgSi$_2$O$_6$ clinopyroxenes (see text). Higher curve takes account of predicted differences in ordering state between 600 and $1300°C$ pyroxenes.

would lead to the enthalpies of mixing at $600°C$ clinopyroxenes being up to 750 J less than those of $1,300°C$ of pyroxenes. Addition of this effect produces a corrected curve (Fig. 4.5) which agrees with the calorimetric data within experimental error.

The fit between calorimetry and experiments could probably be improved a little, but forcing them to fit exactly would not be justified without more data. Additional experiments, involving a redetermination of Holland's $P–X$ diagram at higher temperatures, would show whether the excess free energy of mixing has the kind of temperature dependence discussed here, or if the relationship is more complicated. Clearly, however, bearing in mind the assumptions involved, agreement between calorimetric and phase equilibrium data is already good enough to enable description of mixing properties on this join with considerable confidence. The next step in building towards complex systems is to consider phase relationships involving two binary solid solutions. Before explicitly considering partitioning between such solid solutions, however, it is of interest to look at their combined effects on displaced equilibria.

4.4 Displaced equilibria with two solid solutions

Orthopyroxenes with more than about 65 mol% of the ferrosilite, $Fe_2Si_2O_6$, component are unstable at atmospheric pressure, breaking down to fayalitic olivine and quartz. Since ferrosilite is denser than olivine plus quartz, the effect of increasing pressure must be to stabilize progressively more iron-rich orthopyroxene until some high pressure where pure $Fe_2Si_2O_6$ itself becomes stable.

Bohlen et al. (1980) determined the ferrosilite stability field by reversal of the position of the univariant breakdown reaction:

$$Fe_2Si_2O_6 \rightleftharpoons Fe_2SiO_4 + SiO_2 \qquad (4.9)$$

orthopyroxene   olivine   quartz

Orthopyroxene was found to be stable at about 10.5 ± 0.1 kbar at $700°C$ with reaction pressure increasing to 14.8 kbar at $1,050°C$ (Fig. 4.6). In rocks, of course, the olivine and orthopyroxene are both Fe–Mg solid solutions and the latter contains small amounts of Ca, Mn, Fe^{3+} and Al, too. Therefore, the univariant reaction is not directly applicable to natural systems, although it does provide an important constraint on the free energy of ferrosilite relative to fayalite plus quartz.

In order to approach nature more closely, Bohlen & Boettcher (1981) determined the stabilizing effect of adding $Mg_2Si_2O_6$ component to the orthopyroxene. Their experiments were performed in a piston-cylinder apparatus using an essentially frictionless pressure medium (Ch. 3) with oxygen fugacity controlled close to the iron–wustite buffer in order to minimize Fe^{3+} substitution. Figure 4.6 shows the positions of reversed boundaries for the lower stability limits of orthopyroxene with between five and 20% of the magnesium end-member. Once magnesium is added to the system, it enters both orthopyroxene and olivine, of course, but tends to be concentrated in the former. The pressure shift in orthopyroxene–olivine–quartz equilibrium at temperature T relative to the end-member (ferrosilite) curve may be calculated from a relationship analogous to Equation 4.6:

$$P = P^0 - \frac{RT}{\Delta V^0} \ln \left[\frac{(X_{Fe}^{Ol})^2}{(X_{Fe}^{Opx})^2} \cdot \frac{\gamma_{Fe}^{Ol}}{\gamma_{Fe}^{Opx}} \right]$$

In this expression, X_{Fe}^{Ol} and X_{Fe}^{Opx} are the ratios of $Fe^{2+} : (Fe^{2+} + Mg)$ in olivine and orthopyroxene, respectively. Iron–magnesium partitioning

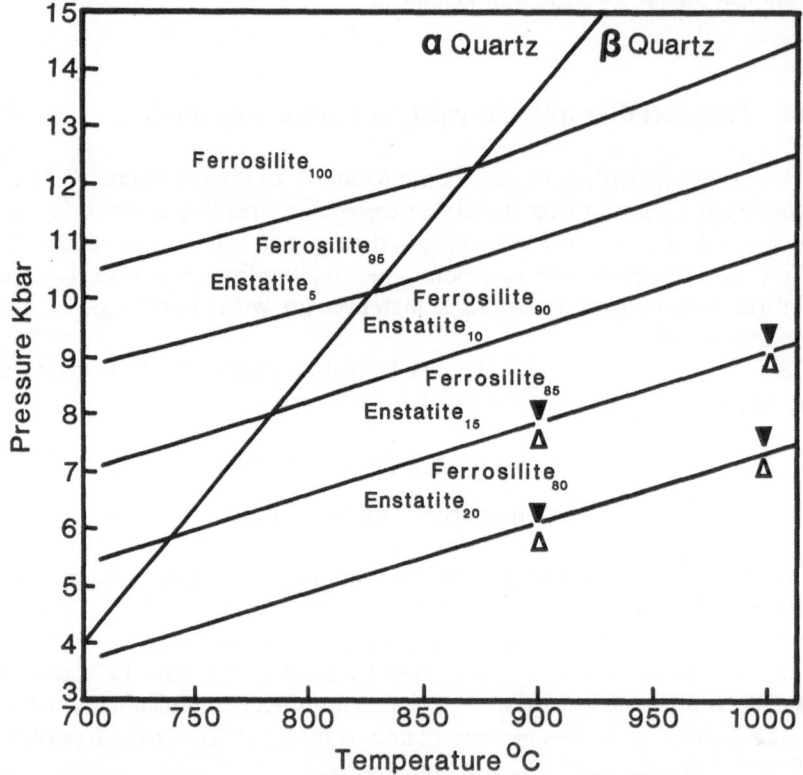

Figure 4.6 Pressure–temperature projection showing relative effects of 5, 10, 15, and 20 mol% MgSiO₃ on the stability of ferrosilite. From Bohlen *et al.* (1980) and Bohlen and Boettcher (1981).

between the two minerals may be characterized by a partition coefficient $K_D = [X_{Fe}^{Ol} \cdot X_{Mg}^{Opx}/(X_{Fe}^{Opx} \cdot X_{Mg}^{Ol})]$ which, from reversed phase compositions, is 2.2 ± 0.3 at $900°C$ when olivine and orthopyroxene are both close to the iron end-member compositions. This K_D, coupled with the pressure equation above, enables complete characterization of the three phase assemblages as a geobarometer for many natural compositions. In addition, Bohlen & Boettcher were able to show from the observed values of P and X that the activity coefficient ratio, $\gamma_{Fe}^{Ol}/\gamma_{Fe}^{Opx}$ is very close to 1.0 in the temperature range 800–1,000°C. Their data thus constrain the relative mixing properties of olivine and orthopyroxene as well as calibrating a useful olivine–orthopyroxene–quartz geobarometer.

4.5 Coexisting solid solutions

As shown above, the fact that natural minerals are solid solutions, while increasing the complexity of their phase relationships, also extends their usefulness for geobarometry and geothermometry. In the case of olivine–orthopyroxene–quartz assemblages, the addition of magnesium extends the pressure range over which three-phase assemblages are stable (in different bulk compositions) from a univariant line to a field which reaches down to atmospheric pressure. Another good example is the two-feldspar geothermometer of Barth (1952). Potassium feldspar and anorthite are stable together over a wide range of P and T, but they have very little mutual solubility. Addition of $NaAlSi_3O_8$ component to the two feldspars complicates the system, of course, but it also provides a way of determining the temperatures of crystallization of coexisting feldspars in rocks. This is because the partitioning of $NaAlSi_3O_8$ between alkali feldspar and plagioclase is strongly temperature-dependent (Barth 1952, Whitney & Stormer 1976).

Many partitioning thermometers rely on exchange equilibria involving magnesium and Fe^{2+} components of coexisting solid solutions. Simple exchange reactions are set up as follows:

$$Fe^{2+} \quad + \quad Mg^{2+} \quad = \quad Fe^{2+} \quad + \quad Mg^{2+}$$

mineral (A) mineral (B) mineral (B) mineral (A)

and the partition coefficient K_D defined as:

$$K_D = \frac{X_{Fe^{2+}}^{B}}{X_{Fe^{2+}}^{A}} \cdot \frac{X_{Mg}^{A}}{X_{Mg}^{B}}$$

where X_{Mg}^{A} is the atomic ratio of magnesium to magnesium plus ferrous iron in mineral A.

If K_D is either much larger or much smaller than 1, then it is generally a strong function of temperature and it may be calibrated as a geothermometer for coexisting natural Fe–Mg solid solutions. Garnet has a very strong affinity for Fe^{2+} relative to Mg, so that most K_D's involving this phase are strongly temperature dependent and have the potential to work as geothermometers, e.g., garnet–biotite, garnet–cordierite, garnet–clinopyroxene and garnet–olivine.

Ferry & Spear (1978) determined the partitioning of Fe and Mg between garnet and biotite at 2 kbar pressure and temperatures of 550 to 800°C. Garnet is extremely sluggish to react under these conditions, while biotite reacts reasonably well. Knowledge of their relative reactivities led Ferry & Spear to use a novel experimental method. Instead of mixing the two phases in roughly equimolar proportions, they used a molar ratio of garnet to biotite of 98 : 2. They synthesized garnets of Fe/Fe + Mg of 0.8 to 0.9 and mixed these with biotites having X_{Fe} of between 0.25 and 1.0. By using an overwhelming quantity of garnet, the garnet composition remained essentially fixed throughout the experiment, even if the initial biotite was relatively magnesian. Thus, the garnet was not required to undergo substantial reaction. At the end of the experiment, the fine-grained product biotites were analyzed with the electron microprobe after dispersing them on a polished diamond surface. The equilibrium biotite compositions could be successfully reversed, even at 550°C, demonstrating the efficacy of the method. Ferry & Spear's calibration of the garnet–biotite geothermometer is most useful at low temperatures when applied to iron-rich, calcium-poor rocks such as metapelites.

In general, because of the deviations of Fe^{2+}–Mg solid solutions from ideal mixing behavior, K_D thermometers are dependent on the Fe : Mg ratios of the coexisting minerals. It is therefore necessary to determine the magnitudes of these compositional effects. O'Neill & Wood (1979) determined the partitioning of Fe^{2+} and Mg between coexisting olivine and garnet as a function of both temperature and composition. As may readily be seen from a plot of $\ln K_D$ ($K_D = X_{Fe}^{Gt} \cdot X_{Mg}^{Ol}/[X_{Fe}^{Ol} \cdot X_{Mg}^{Gt}]$) versus the mole fraction of Mg in olivine (Fig. 4.7), partitioning is strongly composition-dependent. In principle, the observed effect provides considerable information on the properties of the two solid solutions. Consider a set of experiments performed under conditions of constant temperature and pressure with varying Mg : Fe^{2+} ratios of the reactants. The relationship between observed K_D and the equilibrium constant K_a may be written down by adding in the activity coefficients for the two minerals as follows:

$$RT \ln K_a = RT \ln \left[\frac{X_{Fe}^{Gt} X_{Mg}^{Ol}}{X_{Fe}^{Ol} X_{Mg}^{Gt}} \right] + RT \ln \left[\frac{\gamma_{Fe}^{Gt} \gamma_{Mg}^{Ol}}{\gamma_{Fe}^{Ol} \gamma_{Mg}^{Gt}} \right] \qquad (4.10)$$

If olivine and garnet were both ideal, then the γs would all be 1.0 and K_D

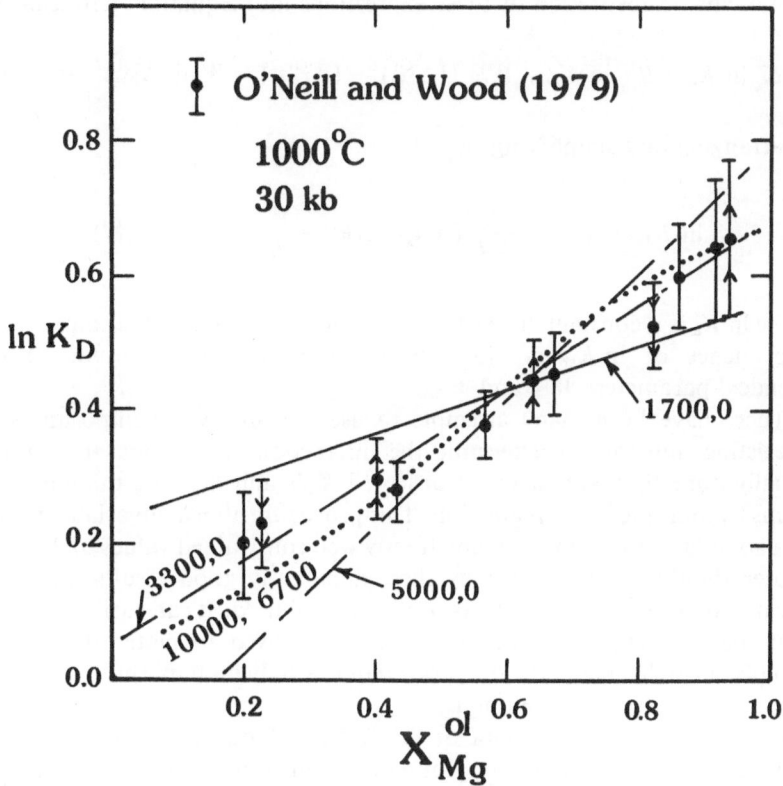

Figure 4.7 Comparison of observed and calculated composition dependence of ln K_D. Uncertainties in data correspond to 1–2 mol% compositional uncertainties. Calculated curves are labeled with assumed Wol and Wgt values, respectively (Eqn. 4.11). Data from O'Neill and Wood (1979).

would be equal to the equilibrium constant K_a and therefore independent of composition at fixed P and T. Obviously this does not apply (Fig. 4.7), so it is necessary to use the next highest approximation, that olivine and garnet each behave as symmetrical regular solutions. In this approximation, the activity coefficients are given by:

$$RT \ln \gamma_{Fe}^{Gt} = W_{Gt}(X_{Mg}^{Gt})^2$$

and

$$RT \ln \gamma_{Mg}^{Ol} = W_{Ol}(X_{Fe}^{Ol})^2$$

etc., where W_{Gt} and W_{Ol} are non-ideal interaction parameters for garnet

63

and olivine. Substitution of these expressions into Equation 4.10 leads to:

$$RT \ln K_a = RT \ln K_D + W_{Gt} [(X_{Mg}^{Gt})^2 - (X_{Fe}^{Gt})^2] - W_{Ol}[(X_{Mg}^{Ol})^2 - (X_{Fe}^{Ol})^2]$$

Rearranging and simplifying, we obtain:

$$\ln K_D = \ln K_a + \frac{W_{Ol}}{RT} (X_{Mg}^{Ol} - X_{Fe}^{Ol}) - \frac{W_{Gt}}{RT} (X_{Mg}^{Gt} - X_{Fe}^{Gt}) \qquad (4.11)$$

Since $\ln K_a$ is a constant at fixed pressure and temperature, the composition dependence of $\ln K_D$ provides information on the magnitude of the nonideal parameters W_{Ol} and W_{Gt}.

There have been many attempts to use the observed compositions of coexisting minerals to determine the interaction parameters W_i. This is usually done by taking a set of observed K_D's and deriving the non-ideal terms by multiple linear regression. This procedure almost invariably results in a good fit to the data, with apparently well constrained values of W_i. The reader should not be fooled by these apparently good results, however, because they are largely artifacts of the method used. It can easily be shown that the absolute values of W cannot be derived to better than about $\pm 6,000$ J by this method. The proof requires only an unprejudiced eyeball and no fancy statistical methods.

Consider the $1,000°C$ data of O'Neill & Wood (1979) for coexisting olivine and garnet. These are plotted as a function of X_{Mg}^{Ol} in Figure 4.7. Note that it is virtually impossible to obtain coexisting mineral compositions to better than 1–2 mol% with the electron microprobe, and that even an uncertainty of this size leads to a substantial uncertainty in $\ln K_D$ (Fig. 4.7). On Figure 4.7, the experimental data are compared with calculated $\ln K_D$ lines obtained using Equation 4.11 for various values of W_{Ol} and W_{Gt}. It can readily be seen that the experimental data cannot really discriminate between $W_{Ol} = 3,330$ J mol, $W_{Gt} = 0$, and $W_{Ol} = 10,000$ J/mol with W_{Gt} of 6,700 J/mol. On the other hand, the experiments do allow discrimination of $W_{Ol} = 5,000$, $W_{Gt} = 0$ from $W_{Ol} = 1,700$, $W_{Gt} = 0$ J/mol. The experimental data do not, therefore, constrain the absolute values of interaction parameters very well. They do, however, provide good information on the difference between olivine and garnet properties, and it appears from Figure 4.7 that $W_{Ol} - W_{Gt}$ is about 3,300 J/mol. The fact that absolute values are poorly constrained is hardly surprising since it can be seen that the W_{Ol} and W_{Gt} terms in Equation 4.11 have exactly the same form. Thus the W values derived from the experiments will be strongly correlated with one another and hence poorly constrained. If, however, the mixing properties of one of the minerals are already known, then partitioning experiments enable those of the other to be derived because the experiments give a precise measure of the difference between the two ($W_{Ol} - W_{Gt}$ in this case).

In order to get absolute values for one of the minerals it is necessary to go back to direct activity–composition determination by the method of displaced equilibrium. Calorimetry can also help provide an answer. Wood & Kleppa (1981) determined the heats of mixing of forsterite–fayalite olivines by solution calorimetry at 700°C. Their results, which are consistent with high-temperature displaced equilibrium measurements, indicate a small positive deviation from ideality of the olivine solid solution. The magnitude of the effect, which is actually asymmetric, corresponds to W_{Ol} of about 4,000 J/mol at the magnesium end of the series. Since the partitioning data require ($W_{Ol} - W_{Gt}$) to be about 3,300 J/mol, the value of W_{Gt} is about +700 J/mol. The result is only reasonably well determined for the Mg-end of the series where O'Neill & Wood made most of their experiments.

4.6 Putting it all together

At several points in this chapter we have emphasized the integration of phase equilibrium and calorimetric data with the goal of building up to the prediction of phase diagrams for natural rock compositions. In recent publications, we have collected together thermodynamic and mixing properties for anhydrous phases in the system Na_2O–FeO–CaO–MgO–Al_2O_3–SiO_2 (Wood & Holloway 1982, 1984, Wood 1987) and have attempted to calculate phase diagrams with them. A good example of their utility can be found in the gabbro to eclogite transformation.

The study by Green & Ringwood (1967) provided an appreciable quantity of data bearing on the stability of garnet and plagioclase in a wide range of mafic compositions. They found that, with increasing pressure, the pyroxene–plagioclase assemblage of basalts is replaced by garnet plagioclase–pyroxene (garnet granulite) and eventually, in the eclogite facies, by garnet–pyroxene–quartz. Because of slow reaction rates, the experiments could only be performed at 1,000°C and above, so that it is necessary to extrapolate the results to lower temperature in order to apply them to conditions of the continental crust.

Figure 4.8 shows a comparison of calculated phase relationships for Green & Ringwood's quartz–tholeiite (A) composition with experimental data for the same composition. The calculations were made with the data base given by Wood (1987). Note that the calculations were based only on calorimetry and simple system experiments and were not constrained in any way to fit the data from the natural system. By any criterion the agreement must be considered exceptional. The calculations provide a lot of information which was not intuitively obvious to Green & Ringwood. In order to

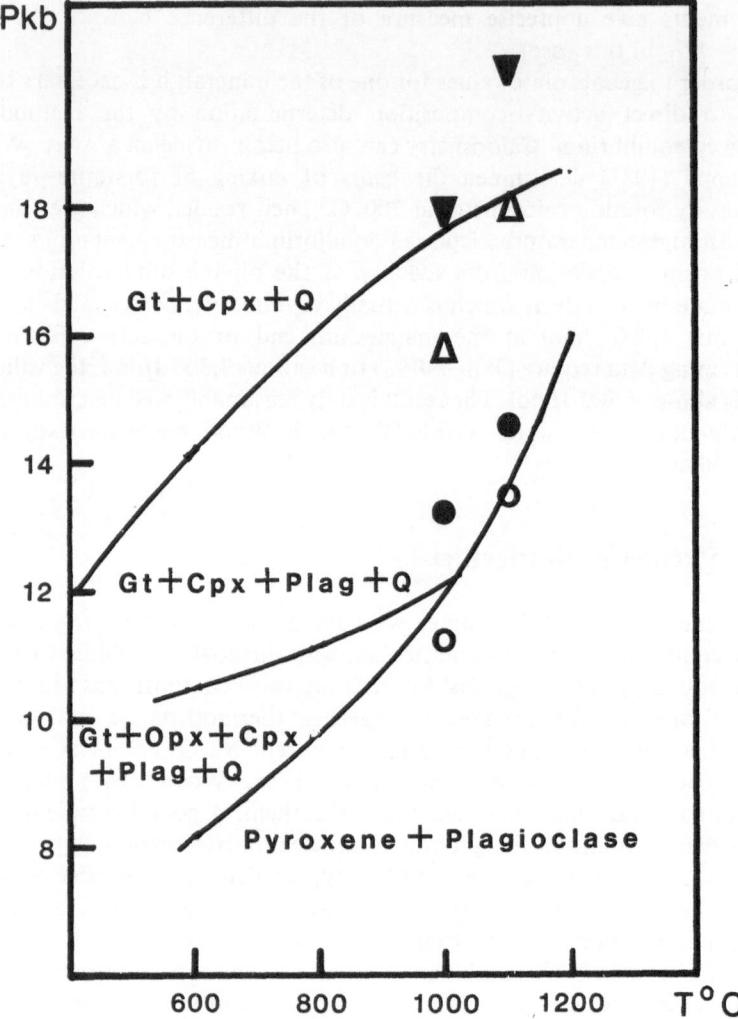

Figure 4.8 Comparison of calculated gabbro-eclogite transition with data from Green and Ringwood (1967). Open circles = Gt absent; filled circles = Gt present; open triangles = plagioclase present; filled triangles = plagioclase absent. Solid lines = (a) calculated appearance of gt (lowest), (b) disappearance of opx, (c) disappearance of plag (highest).

extrapolate their data to lower temperatures, they made the very reasonable assumption that the garnet-in and plagioclase-out curves are parallel to related reactions in simple systems. Based on this assumption, they concluded that eclogite would be the stable mineralogical state of dry basaltic rocks throughout the entire continental crust. However, the calculations demonstrate that the boundaries are actually curved and much flatter than the original extrapolations. The reason for the flattening and curvature is solid solution. The compositions of the minerals change along the garnet-in and plagioclase-out boundaries so that the fixed composition, straight-line assumption is incorrect. Because the calculations reproduce the experiments so well, one can have confidence in concluding that the eclogite stability field is much smaller than was proposed by Green & Ringwood, and that eclogite is not stable under normal crustal conditions.

Should the reader think that our data base is now perfect, and that no further experiments are needed, we should state that some minerals do not fit into the model too well. For example, it was necessary, in performing the calculations shown in Figure 4.8, to make some sweeping assumptions about the properties of garnet. While these work adequately for the quartz–tholeiite composition, the model falls down at very Fe-rich or Ca-rich compositions. A well constrained study of the thermodynamic properties of $Fe_3Al_2Si_3O_{12}$ (almandine) garnet and of its activity in Ca–Mg–Fe solid solutions would greatly help matters.

5 Metamorphic experiments on solid–fluid reactions

5.1 Introduction

A fluid phase generally accompanies the mineralogical and textural changes which take place during progressive metamorphism, and its presence greatly influences the whole metamorphic process. The importance of such a fluid is not, of course, obvious from looking at the finished product of the process, which may be a dense, volatile-free rock with about zero porosity and permeability. There are, however, several reasons for our asserting its importance.

It has been known at least since the work of Barrow (1893, 1912) and Tilley (1925) that progressive metamorphism involves a reduction in the amounts of combined H_2O and CO_2 present in rocks. As metamorphic temperature increases minerals such as chlorite and dolomite, which contain large amounts of H_2O and CO_2, are replaced by, for example, garnet and diopside which have essentially none of these components. Since combined H_2O and CO_2 systematically decreases in amount with increasing temperature, it is reasonable to assume that they are driven off as a separate fluid during the progress of metamorphic reactions. More direct evidence for the presence of a fluid are trapped bubbles of CO_2, H_2O, and gas mixtures which are commonly found in metamorphic minerals (Crawford & Hollister 1986). Careful study of the properties of these "fluid inclusions" has shown that many were trapped under $P-T$ conditions approximating those of the peak of metamorphism. They therefore represent samples of fluids which were present at high P and T during recrystallization. A final piece of evidence concerning fluids is provided by the large scale mass-transfer phenomena which are often associated with metamorphism. The presence of abundant veins and segregations in metamorphic terranes and the mineralization associated with igneous intrusives cannot be interpreted in terms of solid state processes. They require the involvement of a fluid phase capable of carrying substantial amounts of solutes (SiO_2, Ca, Fe, Sn, S, etc.) over considerable distances.

The investigation of solid–fluid reactions is performed in an analogous manner to that described in Chapter 4 for solid–solid reactions, and the best studies have all the ingredients described in Chapter 2. In particular, the need to use well characterized product and reactant phases and the necessity of "reversal" to establish equilibrium boundaries cannot be overemphasized.

Experiments involving solids and fluids have been widely used as the basis for thermodynamic models of metamorphic equilibria. In the most comprehensive studies to date, Helgeson *et al.* (1978) and Berman *et al.* (1985) derived the relative free energies of approximately 70 minerals using a large number of data on different solid–fluid and solid–solid reactions. The essence of the method used to derive free energy data was outlined in Chapter 4.

Consider the dehydration reaction studied by Chatterjee & Johannes (1974):

$$KAl_3Si_3O_{10}(OH)_2 \rightleftharpoons KAlSi_3O_8 + \quad Al_2O_3 \quad + H_2O \qquad (5.1)$$

$$\text{muscovite} \qquad \text{sanidine} \qquad \text{corundum} \qquad \text{fluid}$$

The condition of equilibrium between the four phases involved in the reaction is

$$\Delta G^0 = -RT \ln \left(\frac{a_{H_2O}^{fluid} \cdot a_{KAlSi_3O_8}^{sanidine} \cdot a_{Al_2O_3}^{corundum}}{a_{KAl_3Si_3O_{10}(OH)_2}^{muscovite}} \right) \qquad (5.2)$$

Under the conditions of Chatterjee & Johannes' experiments all of the phases were essentially pure so that, using the conventional standard states of pure phase at the P and T of interest for solids and pure phase at 1 bar and T for fluids, Equation 5.2 becomes:

$$\Delta G^0 = \Delta H_{1,T}^0 - T \Delta S_{1,T}^0 + (P-1)\Delta V_{solids}^0 = -RT \ln f_{H_2O} \qquad (5.3)$$

In Equation (5.3) f_{H_2O} refers to the fugacity of water under the P–T conditions of the experiment and ΔV_{solids}^0 has been assumed independent of pressure and temperature. This assumption is reasonable for solid–fluid reactions at crustal pressures.

The experimental data of Chatterjee and Johannes bearing on Reaction (5.1) are presented in Table 5.1. As can be seen, the reaction was reversed with uncertainties of about $\pm 10°C$ in the P–T range 1–8 kbar and 600–800°C. Each of these reversals constitutes a measurement of ΔG^0 of reaction with an uncertainty, corresponding to the temperature uncertainty, of ± 750 J, approximately. This is about an order of magnitude better than ΔG^0 could be calculated from the calorimetrically measured enthalpies and entropies.

Table 5.1 Hydrothermal runs bearing on the upper thermal stability of 2 M_1-muscovite†

Run no.	P_{H_2O} (kbar)	$T(°C)$	Run time (days)	Condensed phases
Reaction interval at 1 kbar P_{H_2O}, 600–630°C				
18	1	660	40	Kf + C
19	1	640	40	Kf + C
29	1	630	50	Kf + C, Ms decreased
31	1	610	74	Kf + C + Ms; no reaction
30	1	600	81	Kf + C, Ms decreased
Reaction interval at 2 kbar P_{H_2O}, 640–660°C				
11	2	720	20	Kf + C
12	2	700	20	Kf + C
13	2	680	20	Kf + C
21	2	660	30	Kf + C, Ms decreased
28	2	640	42	Kf + C, Ms increased
Reaction interval at 4 kbar P_{H_2O}, 690–710°C				
7	4	740	12	Kf + C
6	4	720	12	Kf + C
26	4	710	12	Kf + C
27	4	700	14	Kf + C + Ms; no reaction
25	4	690	12	Kf + C, Ms increased
15	4	680	12	Ms
Reaction interval at 6 kbar P_{H_2O}, 740–750°C				
8	6	760	6	Kf + C
14	6	750	6	Kf + C, Ms decreased
9	6	740	7	Kf + C, Ms increased
3	6	720	6	Kf + C, Ms strong growth
2	6	700	6	Ms
Reaction interval at 8 kbar P_{H_2O}, 780–800°C				
33	8	800	2.9	Kf + C, Ms decreased
32	8	780	4.7	Ms

†All runs were conducted on a starting material of muscovite bulk composition ($K_2O . 3Al_2O_3 . 6SiO_2$) + excess water. The starting material consisted of synthetic sanidine (Kf) and corundum (C) seeded with 10% by weight of synthetic muscovite (Ms).

Since calorimetrically measured entropies are much better constrained than the enthalpies (Section 3.5) the analysis can be taken a step further (Helgeson *et al.* 1978). Given the molar volumes of the solids, the calorimetric entropy values may be checked for consistency with the $P–T$ slope of the reaction (Section 4.2). By fixing ΔS^0 at the measured (or consistent value) the relative enthalpies of reactants and products can then be determined from the experimental data with greater precision than from solution calorimetry alone. If the enthalpies of formation of a few phases have been measured calorimetrically with good precision, then the experiments enable determination of the enthalpies of formation of others involved in high $P-T$ reactions. In this way, Helgeson *et al.* (1978) built up a large database of enthalpies for minerals which had not been studied calorimetrically or for which the calorimetric data were poor.

Having established several important reasons for studying solid–fluid equilibria we will turn now to some of the practical aspects of their investigation.

5.2 Fluid generation and control

5.2.1 The composition of fluids in the crust

The important constituents of fluids in the Earth's crust are H_2O and CO_2, or, in some cases, H_2O and CH_4. Except in volcanic gases, CO and SO_2 are present in only minor amounts. Because they are closely linked to H_2O, O_2 and H_2 are also important, although O_2 is never present in more than vanshingly small amounts. In addition to these molecular species, ionized acids, bases, and salts may be present in major amounts under lower T conditions ($<400°C$), but become less important at higher T (see Section 5.4.3). The six important species, $H_2O, CO_2, CH_4, H_2, O_2$, and CO are contained in the C–O–H system shown in Figure 5.1.

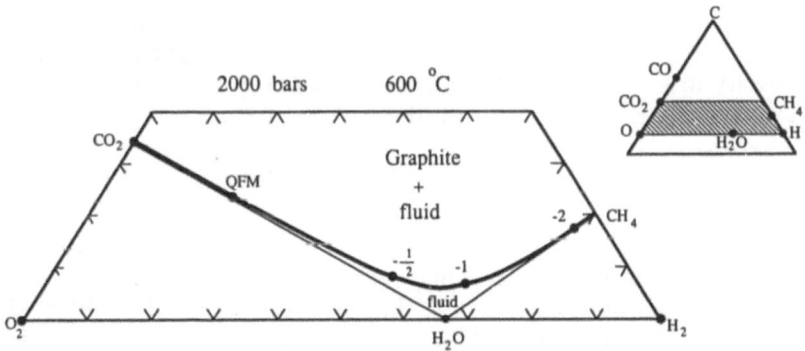

Figure 5.1 The graphite saturation surface at 2,000 bar and 600°C. The heavy line represents fluid compositions in equilibrium with graphite. Fluids lying below the line are undersaturated in graphite. The point labeled "QFM" has an f_{O_2} equal to that of the QFM buffer. Numbers show f_{O_2} in $\log_{10}$ units relative to QFM.

5.2.2 The C–O–H system

The possibilities and problems of generating fluids in experiments at high $P-T$ conditions may be understood with reference to the C–O–H system.

In considering the fluid we assume that there is no significant chemical interaction between it and the silicate portion of the overall experimental system. In particular we ignore changes in fluid composition caused by dissolution of silicate constituents in the fluid.

At constant P and T, the phase rule requires that the variance of the system, $f = c - p = 3 - p$. If only a homogeneous fluid is present then $p = 1$ so $f = 2$. This means that two variables must be fixed in order to fix the composition, and hence the activities, of the fluid species. In most experiments one is interested in fixing H_2O or CO_2 activity, or both.

5.2.3   The bulk composition method

The most common variables to fix are H_2 activity and the $C : O$ ratio in the bulk fluid. H_2O-CO_2 fluids may be generated from mixtures of silver oxalate [$Ag_2C_2O_4$] or anhydrous oxalic acid [$H_2C_2O_4$] or oxalic acid dihydrate [$H_2C_2O_42H_2O$], and H_2O (Holloway et al. 1968; Holloway & Reese 1974). At temperatures above about $200°C$ the oxalates decompose to form CO_2. The composition of the generated fluid depends on the bulk composition loaded into the capsule and on the H_2 activity. The variation in fluid speciation as a function of H_2 activity is shown in Figure 5.2. Note that the fluid is a nearly binary mixture of CO_2 and H_2O over a wide range of H_2 activities commonly found in high pressure vessels. Because of this wide latitude of allowed H_2 activities, many experiments have been done without explicitly controlling H_2 activity, a practice that is acceptable only if the sample does not contain iron. In experiments of this type it is especially necessary to know the fluid composition. However, the fluid composition may be changed by reactions with the sample, such as:

$$Mg(OH)_2 \rightleftharpoons MgO + H_2O \tag{5.4}$$

The liberated H_2O is added to the fluid, thereby increasing the $H_2O : CO_2$ ratio and the H_2O activity. There are three ways to know the fluid composition in experiments such as this. One way is to analyze the fluid at the end of the experiment, a second method is to calculate the fluid composition from a knowledge of the masses of fluid and sample placed in the capsule, and a third technique is to make the ratio of fluid to sample so large that the fluid composition is not significantly changed by reactions in the sample. The effect of varying proportions of fluid to sample mass is illustrated in Figure 5.3 for the case of mixtures of oxalic acid dihydrate and a silicate melt. In this case the fluid and melt compositions each depend on the fluid : sample mass ratio, but it is possible to calculate the final result. It can be seen from the figure that for fluid : sample mass ratios greater than about $2 : 1$ the actual and "theoretical" fluid compositions are essentially equal.

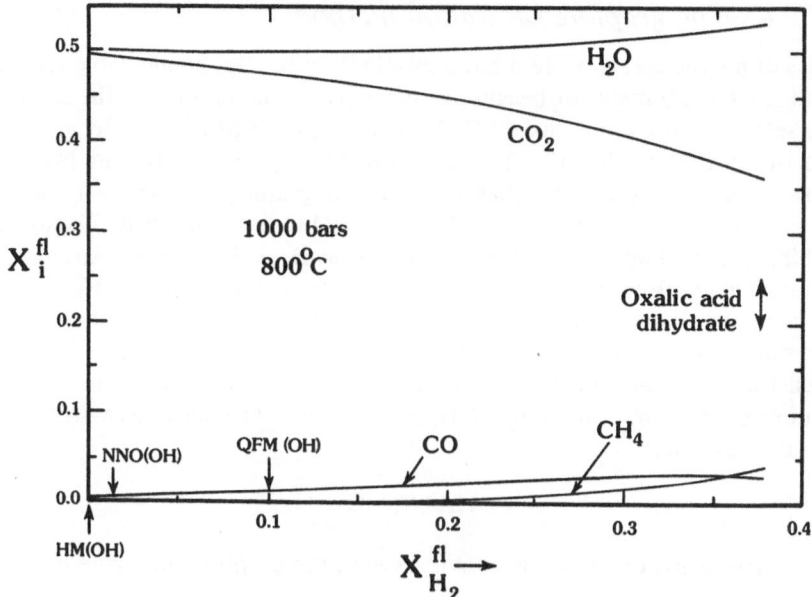

Figure 5.2 Fluid composition as a function of X_{H_2} for an oxalic acid dihydrate initial bulk composition (shown on the far right-hand side of the diagram). Values of $f_{H_2}^{fl}$ corresponding to hematite–magnetite, Ni–NiO, and QFM coexisting with H_2O are shown.

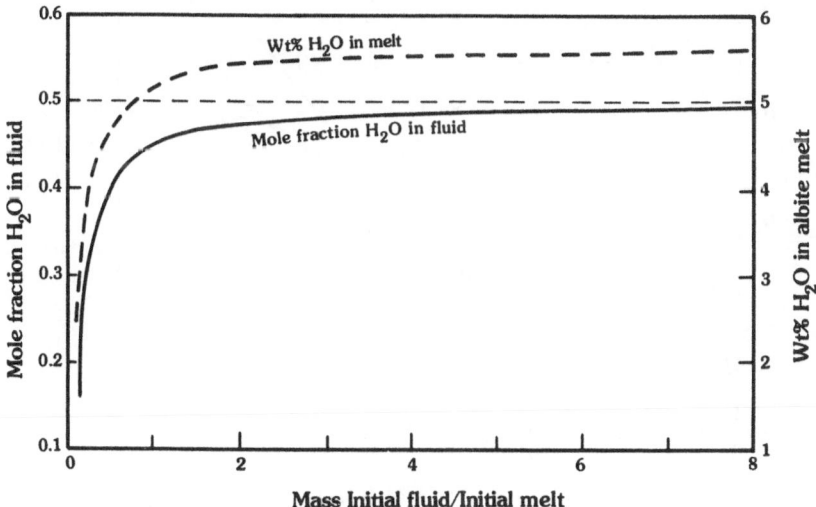

Figure 5.3 Variation in the mole fraction of H_2O in an $H_2O–CO_2$ fluid phase coexisting with silicate melt of albite composition as a function of the initial fluid:melt ratio with fluid provided by oxalic acid dihydrate.

5.2.4 The graphite saturation method

Use of a large excess of fluid has limitations, either due to available volume in high P equipment, or because of incongruent dissolution of the sample. A second technique to fix C–O–H fluid compositions is to add graphite to the system so that the fluid is saturated in graphite. In this instance the system contains two phases (fluid and graphite) so the variance is $f = 3 - 2 = 1$, and only one additional variable need be fixed. Eugster & Skippen (1967) describe techniques for fixing H_2 activity using external H_2 buffers in graphite saturated systems. (Another method of fixing H_2 activity is with the Shaw hydrogen membrane described in the Appendix). Fluid compositions in equilibrium with graphite are shown in Figure 5.4 as a function of H_2 activity for various external buffers. One common external buffer consists of a mixture of H_2O, Ni, and NiO which react to fix H_2 activity as follows:

$$Ni + H_2O \rightleftharpoons NiO + H_2 \qquad (5.5)$$

Another means of fixing H_2 activity is with the graphite–methane buffer:

$$C + 2H_2 = CH_4 \qquad (5.6)$$

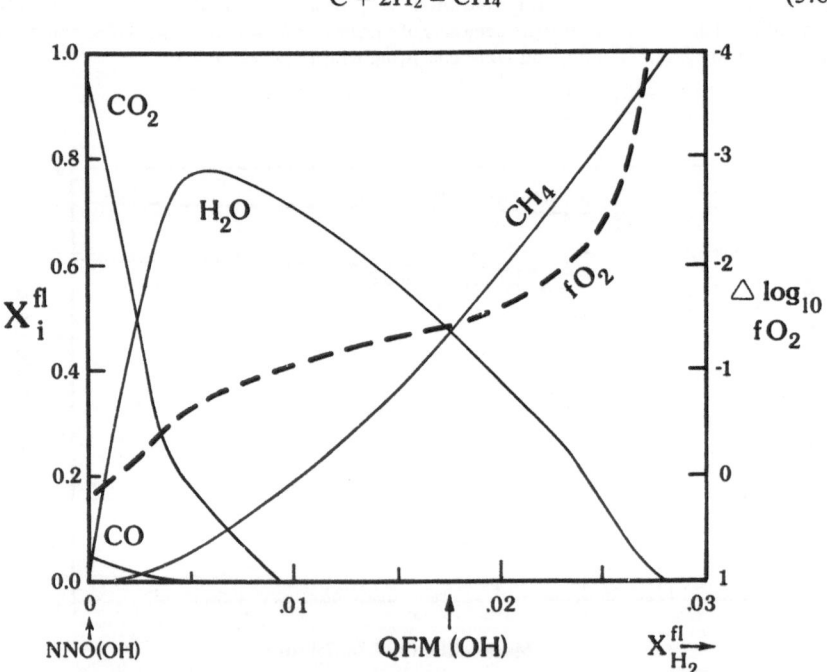

Figure 5.4 Graphite-saturated fluid compositions at 2,000 bar, 600°C. Solid lines refer to mole fractions of major species (left-hand scale). Dashed line refers to relationship between mole fraction H_2 (bottom) and Log f_{O_2} relative to the NNO buffer (right hand scale).

which has been used in cold-seal vessels with methane as the pressure medium (Eugster & Skippen 1967). The graphite saturation technique works well provided the experimental T is sufficiently high for graphite to equilibrate (probably $> 600°C$). Because all C–O–H fluid activities are fixed in this situation, oxygen fugacity and H_2O activity cannot be independently varied. This covariance limits the usefulness of the technique for iron-bearing samples.

5.3 Devolatilization experiments

An extremely well planned example of mixed-volatile experiments is provided by Skippen (1971) in a study of reactions in siliceous marbles. Progressive metamorphism of such rocks, which approximate the chemical system $CaO–MgO–SiO_2–CO_2–H_2O$, produces a large number of different solid phases. Many stable assemblages are, however, mixtures of three or more of the minerals: quartz–calcite–dolomite–talc–tremolite–diopside–forsterite–enstatite, which coexist with mixed $CO_2–H_2O$ fluids. Skippen showed that there are 49 possible reactions linking these eight minerals and the fluid phase. Obviously a complete study of all of them would daunt even the most vigorous and enthusiastic experimentalist. Because of thermodynamic relationships, however, Skippen was able to show that data for all 49 reactions could be derived from results on only five of them. An outline of the reasoning follows.

Skippen obtained equilibrium data for the five reactions:

$$Mg_3Si_4O_{10}(OH)_2 \rightleftharpoons 3MgSiO_3 + SiO_2 + H_2O \qquad (S.1)$$

talc   enstatite   quartz   fluid

$$Mg_3Si_4O_{10}(OH)_2 + 3CaCO_3 + 2SiO_2 \rightleftharpoons 3CaMgSi_2O_6 + 3CO_2 + H_2O \quad (S.2)$$

talc   calcite   quartz   diopside   fluid   fluid

$$Ca_2Mg_5Si_8O_{22}(OH)_2 \rightleftharpoons 2CaMgSi_2O_6 \rightleftharpoons 3MgSiO_3 + SiO_2 + H_2O \quad (S.3)$$

tremolite   diopside   enstatite   quartz   fluid

$$3CaMg(CO_3)_2 + 4SiO_2 + H_2O \rightleftharpoons Mg_3Si_4O_{10}(OH)_2 + 3CaCO_3 + 3CO_2$$

dolomite   quartz   fluid   talc   calcite   fluid

$$(S.4)$$

$$4Mg_3Si_4O_{10}(OH)_2 + 5CaMg(CO_3)_2$$

talc   dolomite

$$\rightleftharpoons 5CaMgSi_2O_6 + 6Mg_2SiO_4 + 10CO_2 + 4H_2O \quad (S.5)$$

diopside   forsterite   fluid   fluid

Experimentally Skippen used mixtures of reactant and product minerals and established the direction of reaction using optical and X-ray methods. Fluid generation and control was performed by the double-capsule technique with Ni–NiO or QFM buffers outside and graphite inside the H_2-permeable capsule.

The positions of the 44 undetermined reactions are obtained in the following way. For Reaction S.1 the equilibrium condition (Eqn. 5.3) may be rearranged to give an expression of form:

$$\log K_1 = \log f_{H_2O} = \frac{A_1}{T} + B_1 + \frac{C_1(P-1)}{T}$$

where A, B and C are related to ΔH^0, ΔS^0 and ΔV^0_{solids}, respectively, and a conversion to base 10 logarithm has been made for convenience. Similarly for Reaction S.2 we have, assuming that CO_2 and H_2O mix ideally:

$$\log K_2 = 3 \log f_{CO_2} \cdot X_{CO_2} + \log f_{H_2O} \cdot X_{H_2O} = \frac{A_2}{T} + B_2 + \frac{C_2(P-1)}{T}$$

Subtracting $\frac{1}{3}$ of Reaction S.1 from $\frac{1}{3}$ of Reaction S.2 gives the reaction:

$$CaCO_3 + SiO_2 + MgSiO_3 \rightleftharpoons CaMgSi_2O_6 + CO_2 \quad (S.6)$$

calcite   quartz   enstatite   diopside   fluid

Since free energy and log K data may be added and subtracted in exactly the same way, we have, for Reaction S.6

$$\log K_6 = \tfrac{1}{3} \log K_2 - \tfrac{1}{3} \log K_1 = \log f_{CO_2} \cdot X_{CO_2}$$

$$= \frac{A_2 - A_1}{3T} + \left(\frac{B_2 - B_1}{3}\right) + \frac{(C_2 - C_1)(P-1)}{3T}$$

All of the other 43 equilibria may be derived by analogous combination methods.

Skippen's approach provides thermodynamic data for all possible equilibria linking the eight solid phases of interest with a minimum amount of experimental effort. There are, however, important reasons for performing more experiments than the absolute minimum. One obvious reason is that,

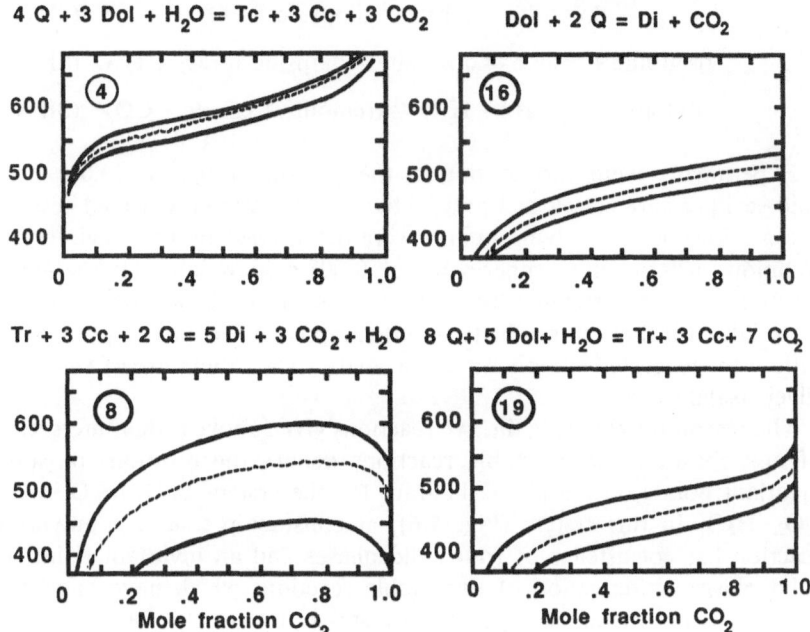

Figure 5.5 Experimental data on Reaction (4) (top left) and calculated positions of (16), (8), and (19) from Skippen (1974). Upper and lower temperature uncertainties are indicated by the solid lines.

when experimental errors are propagated into some of the derived equilibria, the data for the latter became so uncertain that they are almost valueless.

Figure 5.5 shows the positions of a number of calculated reactions at 2 kbar total pressure on plots of temperature versus mole fraction of CO_2 (X_{CO_2}) in the fluid phase (Skippen 1974). As can be seen, the experimentally determined Reaction 4 is very well constrained, as are some of the derived reactions (16 and 19, for example), but several of the latter have very large uncertainties. These uncertainties seriously hamper any attempt to construct the correct phase diagram for the system $CaO-MgO-SiO_2-CO_2-H_2O$. Determination of much of the phase diagram requires additional experimental data.

Slaughter *et al.* (1975) performed a more restricted study than that of Skippen (1971), but one which greatly improves the determination of part of the stable phase diagram. They performed experimental determinations of the positions of the reactions:

$$\text{talc} + \text{calcite} + \text{quartz} \rightleftharpoons \text{tremolite} + CO_2 + H_2O \quad \text{(SKW 1)}$$

$$\text{tremolite} + \text{calcite} + \text{quartz} \rightleftharpoons \text{diopside} + CO_2 + H_2O \quad \text{(SKW 2)}$$

$$\text{dolomite} + \text{quartz} \rightleftharpoons \text{diopside} + CO_2 \qquad\qquad (\text{SKW 3})$$

$$\text{tremolite} + \text{calcite} \rightleftharpoons \text{dolomite} + \text{diopside} + CO_2 + H_2O \quad (\text{SKW 4})$$

$$\text{dolomite} + \text{quartz} + H_2O \rightleftharpoons \text{tremolite} + \text{calcite} + CO_2 \quad (\text{SKW 5})$$

Experiments were carried out in the pressure range 1–5 kbar using cold-seal pressure vessels at 1 and 2 kbar and an internally heated vessel at 5 kbar. Reactions involving quartz were determined by the single crystal technique using quartz spheres about 0.008 g in weight. Quartz-absent reactions were investigated by comparing X-ray peak heights in starting materials with those in products. Fluid compositions of desired $CO_2 : H_2O$ ratio were generated either with oxalic acid or with a mixture of water and silver oxalate.

The reason for choosing the five reactions given above is that intersection of appropriately chosen stable reactions defines the positions of stable invariant points on the phase diagram for the system $CaO–MgO–SiO_2–CO_2–H_2O$. In this system (Fig. 5.6), at constant pressure, a univariant reaction has coexistence of four solid phases and an invariant point five solid phases. Intersection of two stable reactions which have one phase different from one another generates a stable invariant point from which three more univariant reactions radiate. Of the reactions studied by Slaughter *et al.* (1975), SKW 1 and SKW 5 intersect at invariant point B and SKW 2, SKW 3, and SKW 5 at invariant point A. (Note that reaction SKW

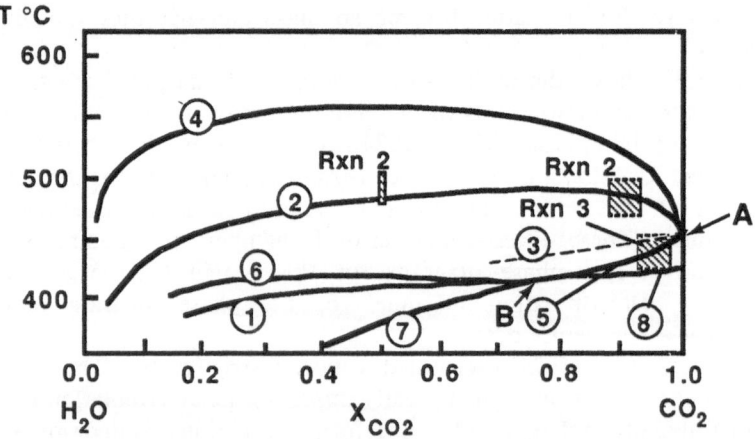

Figure 5.6 Temperature–mole fraction CO_2 plot at 1,000 bar from Slaughter *et al.* (1975). All reactions entering invariant points A and B are shown. The curves are calculated by extrapolations of experimentally determined points with corrections for non-ideality in the fluid phase. The locations and uncertainties of experimental brackets on Rxn.2 and Rxn.3 are shown as hatched areas.

3 is degenerate, with only three solid phases because diopside, dolomite, and quartz are chemographically coplanar). The positions of the additional reactions are determined by Schreinemakers rules and their slopes on the $T - X_{CO_2}$ diagrams obtained from the equilibrium constants as used by Skippen (1971). In this study additional reactions are determined (Fig. 5.6), but they are better known than many of those discussed by Skippen because their positions depend on the intersections of two or more well determined reactions.

5.4 Mineral–fluid interactions

The importance of the fluid phase as an agent of mass transfer during metamorphism was alluded to in the introduction to this chapter. Generation of veins, dominantly of quartz, during progressive regional metamorphism, is usually ascribed to fluid movement along fractures and precipitation of silica from solution during continued cooling and decompression. This interpretation is confirmed by observations in geothermal systems in which groundwaters are heated by magma bodies which come close to the Earth's surface. Temperatures in the heated groundwaters are sometimes in excess of $300°C$, with the water kept from boiling by the higher than atmospheric pressure. When such solutions are sampled in wells or springs they are found to contain much higher concentrations of SiO_2 than would be possible at $25°C$. At $300°C$ concentrations of about 1,000 ppm SiO_2 are to be found whereas quartz-saturated solutions at $25°C$ would contain only 6 ppm SiO_2. The implication is that high temperature water is a much more effective solvent for quartz than water at the Earth's surface. A practical application of the observation is that the silica contents of geothermal waters may be used to estimate the temperatures in the reservoir from which they come (e.g., Fournier & Rowe 1966, Ragnarsdottir & Walther 1983).

Since migrating fluid has the potential to carry along substantial amounts of solutes, geological evidence of mass transfer might, in principle, be used to deduce how much fluid has passed through a particular rock or around an igneous pluton. In order to make such calculations, however, it is necessary to perform solubility experiments in which the equilibrium solute concentrations are measured at fixed activity. The simplest kind of experiment is one in which a single phase such as quartz $(SiO_2)_{solid}$ dissolves congruently to form a silica species in solution.

5.4.1 Congruent dissolution

Anderson & Burnham (1965) determined the solubility of quartz in water in the $P–T$ range 1–10 kbar and $500–900°C$. The experiments were performed by putting crushed quartz in a perforated gold capsule which was then weighed and placed inside a larger gold capsule together with between 5 and 20 grams of water. The large capsule was sealed and then run in an internally heated pressure vessel at the desired run conditions for times up to a week. On quenching, the capsule was cut open and the inner capsule weighed, dried, and then reweighed. The weight difference between dry capsule before the run and after the run gave the amount of silica dissolved by the aqueous fluid. The after-run drying procedure enabled correction for the small amount of water trapped inside the inner capsule.

The results of Anderson & Burnham's solubility experiments are shown in Figure 5.7. As can be seen, the solubility of quartz increases substantially

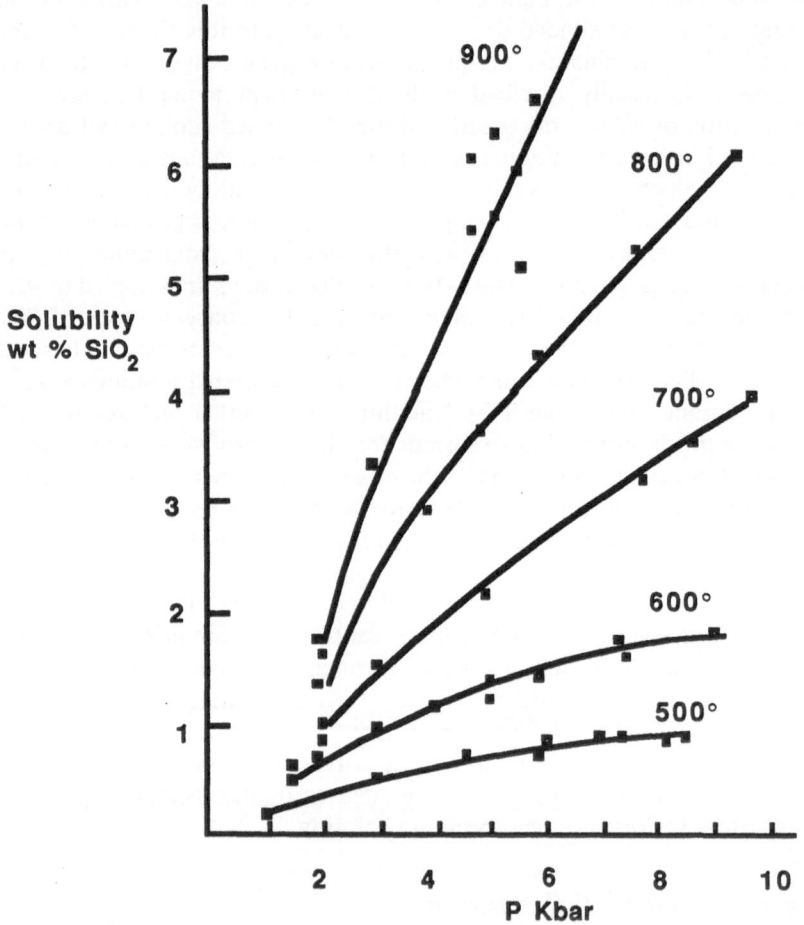

Figure 5.7 Solubilities of quartz in H_2O shown as isotherms (°C) as a function of pressure. From Anderson and Burnham (1967).

with increasing temperature, confirming the potential for SiO_2 transfer by migrating fluids during metamorphism. For comparison, quartz-saturated silica concentrations on the order of 50,000 ppm (6%) are observed at $900°C$, 10,000 ppm at $600°C$, 1,000 ppm at $300°C$ and 6 ppm at $25°C$.

Anderson & Burnham were not able to reverse their measurements with the double-capsule technique they used. Therefore, in order to establish proximity to equilibrium, they performed runs for longer and longer time to find out at what point the rate of increase of silica concentration with time approached zero. They then ran for times long enough to reach a "plateau" region of negligible change in concentration with time.

Although solubility data collected with pure water place some constraints on metamorphic fluid compositions, it is apparent from Section 5.2 that many (if not all) metamorphic fluids contain substantial amounts of components other than H_2O and SiO_2. One of the most important additional components is, of course, CO_2 and quartz solubility in pure CO_2 is virtually zero. In order to extrapolate from data for pure H_2O fluid to fluids containing CO_2, CH_4, $NaCl$, etc., it is necessary to determine activity–composition relations for the SiO_2 species in mixed fluids. Solubility measurements can, in principle, provide much of this information if experiments are performed which enable the dominant SiO_2 species in the fluid to be identified.

The solubility of quartz in water is so high that it must involve formation of an aqueous complex with the solvent. If this were not so, and unsolvated SiO_2 were the dominant species, solubilities in H_2O and CO_2 fluids would be comparable instead of dramatically different.

Bearing in mind the formation of a complex, the quartz dissolution reaction is generally written as:

$$SiO_2 + nH_2O \rightleftharpoons SiO_2 . nH_2O \qquad (5.7)$$

$$\text{quartz} \qquad \text{fluid} \qquad\qquad \text{fluid}$$

Under fixed conditions of pressure and temperature the equilibrium constant is:

$$K = \frac{a_{SiO_2} . nH_2O}{a_{SiO_2}^{quartz} . a_{H_2O}^{n}}$$

With quartz pure and in its standard state (activity equal to 1.0), the nature of the aqueous complex may be deduced by determining quartz solubility under conditions of variable activity of water:

$$\text{d} \log a_{SiO_2} . nH_2O = n \, \text{d} \log a_{H_2O} \qquad (5.8)$$

Therefore a plot of $\log a_{SiO_2} . nH_2O$ versus $\log a_{H_2O}$ at fixed P and T gives a slope of n which establishes the nature of the dominant species. Furthermore, knowledge of the relations between solubility and a_{H_2O} enable extrapolation of data on pure water to impure H_2O–CO_2–CH_4, etc., fluids with variable water activity provided no other silica species becomes important in the mixed fluids.

Although there have been several studies of the effects of varying a_{H_2O} on quartz solubility, we are going to emphasize those of Shettel (1974) and Walther & Orville (1983) as illustrating the types of information which can be obtained. Shettel used the same technique as Anderson & Burnham

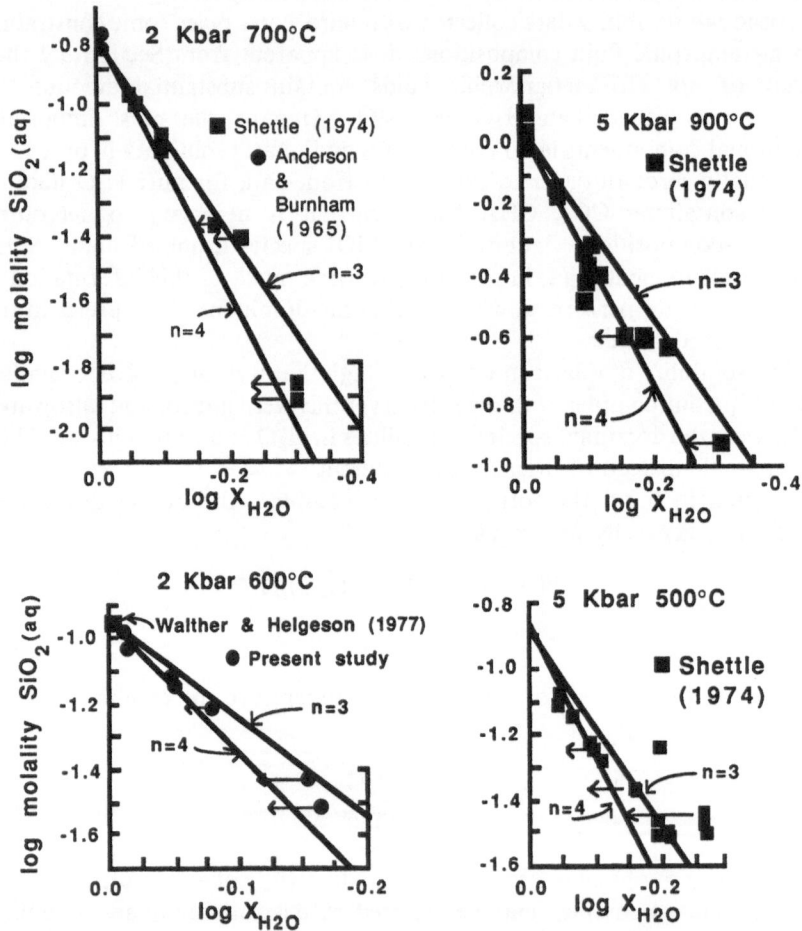

Figure 5.8 Constant pressure and temperature plots for quartz solubility determinations in CO_2–H_2O mixtures for a variety of pressures and temperatures. Lines marked $n = 2$, 3, or 4 give the predicted solubility behavior for these hydration states of aqueous silica. From Walther and Orville (1983). Endpoints of arrows indicate, for representative points, how the data shift if the abscissa is changed from log X_{H_2O} to log a_{H_2O}

82

(1965) and determined the solubility of quartz in $H_2O–CO_2$ mixtures at several pressures and temperatures. Walther & Orville worked with a large volume (35 ml) hydrothermal vessel containing about 5 g of coarse-grained quartz crystals. Their apparatus has capillary tubes passing into the reaction chamber, and the fluid can be sampled at the pressure and temperature of the experiment by opening a valve on one of the tubes. Furthermore, new fluid can be injected into the vessel through the capillaries while the run is at high pressure and temperature. This facilitates the reversing of solubility data because the fluid compositions can be adjusted from old to new values and back again while the run is continuing.

Some of the solubility data of Shettel (1974) and Walther & Orville (1983) are shown in Figure 5.8 as plots of log molality of SiO_2 in the fluid versus log mole fraction of H_2O. The implicit assumption in this kind of diagram is that the activity of SiO_2 in the fluid is proportional to its molality and that H_2O mixes ideally with the other fluid species. As can be seen from Figure 5.8, the solubility–a_{H_2O} relations are consistent with a hydration number $n = 4$ for the aqueous silica complex. The agreement with $n = 4$ is particularly good if the nonideal interaction of CO_2 and H_2O in the fluid phase is included so that a_{H_2O} becomes greater than X_{H_2O} (Fig. 5.8). Similar experiments on H_2O–argon mixes also yield an $n = 4$ (Walther & Orville 1983). It appears, therefore, that the dominant silica species in supercritical fluids is $SiO_2 . 4H_2O$ and that quartz solubility in pure H_2O can be extrapolated to mixed fluids by taking account of how the activity of water varies as a function of fluid composition:

$$d \log m^{fl}_{SiO_2} = 4 \ d \log \ a^{fl}_{H_2O}$$

where $m^{fl}_{SiO_2}$ is the molality of SiO_2 in the fluid phase.

5.4.2 Incongruent dissolution

Most minerals do not dissolve in the simple congruent manner outlined above for quartz. Generally, one or more of their components dissolves preferentially leaving behind a residue enriched in the less soluble species. Feldspars are a classic example. When contacted by aqueous solutions at low temperature, the alkalies are preferentially removed leaving an alumina-enriched material which transforms into kaolinite [$Al_2Si_2O_5(OH)_4$], gibbsite [$Al(OH)_3$], or smectite clays.

The incongruent dissolution relationships of most silicates have been the subject of considerable experimental study because they provide a means of deducing the relative amounts of nonvolatile species (alkalies, Ca, Mg, etc.) in metamorphic fluids. Before considering these experiments in detail, however, it is instructive to look at the use of solubility experiments to determine the relative free energies of minerals other than quartz at high P and T. As an example we will use the study of Hemley et al. (1977).

Several dehydration reactions relevant to the progressive metamorphism of ultrabasic rocks may, in the $MgO–SiO_2–H_2O$ system, be represented as:

$$4Mg_2SiO_4 + 9Mg_3Si_4O_{10}(OH)_2 \rightleftharpoons 5Mg_7Si_8O_{22}(OH)_2 + 4H_2O \quad (H.1)$$

| forsterite | talc | anthophyllite | water |

$$7 \text{ talc} \rightleftharpoons 3 \text{ anthophyllite} + 4 \text{ quartz} + 4 \text{ } H_2O \qquad (H.2)$$

$$\text{anthophyllite} + \text{forsterite} \rightleftharpoons 9 \text{ } MgSiO_3 \text{ (enstatite)} + H_2O \qquad (H.3)$$

$$\text{anthophyllite} \rightleftharpoons 7 \text{ enstatite} + \text{quartz} + H_2O \qquad (H.4)$$

Although these reactions can be studied by conventional hydrothermal techniques, an elegant approach used by Hemley et al. (1977) can give as good results with relatively short run times. The method makes used of the fact that any pair of the minerals of interest, (talc, enstatite, forsterite, and anthophyllite) buffers the silica content of the aqueous fluid in equilibrium with it, e.g.,

$$2MgSiO_3 + 4H_2O \rightleftharpoons Mg_2SiO_4 + SiO_2 . 4H_2O \qquad (A)$$

$$Mg_3Si_4O_{10}(OH)_2 + 3H_2O \rightleftharpoons 3MgSiO_3 + SiO_2 . 4H_2O \qquad (B)$$

Since, under the conditions of these experiments (1 kbar and 300–720°C) the fluid is essentially binary $H_2O–SiO_2$ with very small amounts of Mg, the equilibrium silica molality provides a measure of the relative free energies of the two minerals. Considering Reaction A, coexistence of forsterite and enstatite at equilibrium has the condition:

$$\Delta G_{P,T} = 0 = G^0_{Mg_2SiO_4} - 2G^0_{MgSiO_3} - 4G^0_{H_2O} + G^{obs}_{SiO_2.4H_2O}$$

The free energy of (essentially) pure H_2O is known, as is the partial molar free energy of the aqueous silica complex. The latter may be referred to quartz by considering the difference between the observed molal concentration and that obtained at quartz saturation under the same $P-T$ conditions:

$$G^{obs}_{SiO_2.4H_2O} - G_{Qz} = RT \ln \left(\frac{M^{obs}_{SiO_2.4H_2O}}{M^{Q-Sat}_{SiO_2.4H_2O}} \right)$$

Hence the free energy difference between forsterite and 2 enstatite under the $P-T$ conditions of the experiment is determined.

Hemley et al. (1977) used this approach to obtain the relative free energies of a number of minerals in the $MgO–SiO_2–H_2O$ system. The experiments were performed in standard cold-seal vessels with the charges sealed in platinum capsules. The latter contained between 10 and 50 mg of solid matter and between 0.6 and 1.5 ml of solutions with variable amounts of aqueous silica. The approach to equilibrium was confirmed by showing that solutions with low and high silica contents approached the same value in contact with a specific assemblage. Runs were made for 1 to 60 days, rapidly quenched, capsules opened, and the solution diluted and filtered. The silica contents of the resultant solutions were then determined spectrophotometrically.

Some of the results which Hemley *et al.* used to extract free energy data for talc, anthophyllite, enstatite, and forsterite are shown in Figure 5.9. In addition to the free energy data, the equilibrium temperatures of the dehydration reaction (H.1–H.4) at 1 kbar are well fixed by the points at which any three minerals coexist. For example, the equilibrium temperature of H.1 is fixed at $636 \pm 7°C$ by the intersection of talc–anthophyllite and anthophyllite–forsterite equilibria. Similarly, we have for H.2 (talc–anthophyllite–quartz) $675 \pm 7°C$, for H.3 $670 \pm 7°C$, and for H.4 $730°C$, approximately.

The experiments of Hemley *et al.* were performed in essentially pure water, and the concentrations of dissolved Mg were much less than those of SiO_2. In order to dissolve appreciable amounts of Mg, Ca, and Fe into aqueous fluids it is necessary that they be much more acid than the near-neutral fluids used in the $MgO–SiO_2–H_2O$ experiments. At low

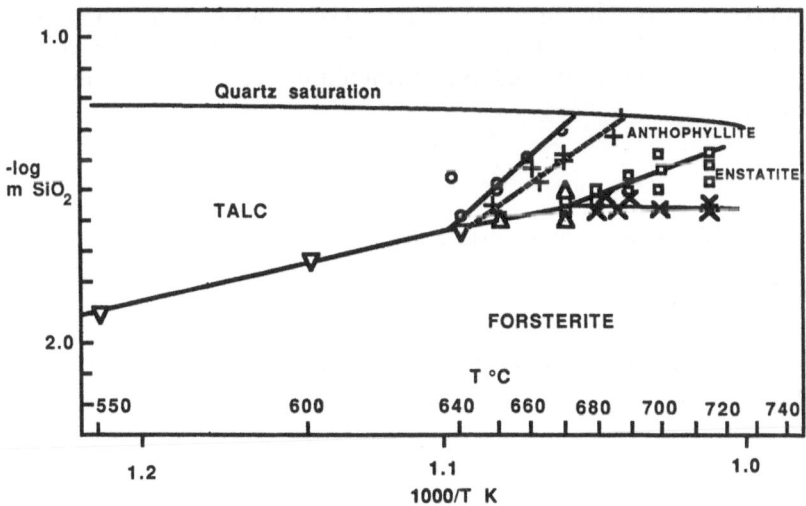

Figure 5.9 Derivation of anthophyllite stability relations at 1 kbar from the variation of silica concentration in solution. From Hemley *et al.* (1977). ∇ talc–forsterite, △ anthophyllite–forsterite, × enstatite–forsterite, □ anthophyllite–enstatite, + talc–enstatite, ○ talc–anthophyllite.

85

temperatures dissolution of silicates in acid solutions may be represented by equilibria such as:

$$MgSiO_3 + 2H^+ + 3H_2O \rightleftharpoons Mg^{2+} + SiO_2 . 4H_2O$$

At temperatures above about 250°C, however, ionized species such as Mg^{2+} start to complex significantly with the anions present in solution. In the metamorphic temperature range it is therefore more appropriate to consider dissolution equilibria such as:

$$MgSiO_3 + 2HCl + 3H_2O \rightleftharpoons MgCl_2 + SiO_2 . 4H_2O \qquad (5.9)$$

rather than those involving the ionized, dissociated species.

Analyses of geothermal fluids indicate that the dominant acid causing silicate dissolution in near-surface systems are HCl and H_2SO_4. Although metamorphic fluids are difficult to sample directly, it is known from analysis of fluid inclusions (Poty *et al.* 1974) that chloride, dominantly as NaCl, is an important constituent of aqueous high-temperature fluids. Therefore it is appropriate to investigate mineral–fluid equilibria and mass transfer in metamorphic systems in terms of chloride species such as those used in Reaction 5.9. To this end, Frantz & Eugster (1973) developed a double capsule method to buffer HCl fugacity in high $P-T$ solution experiments.

The double capsule buffering technique for f_{HCl} is shown in Figure 5.10a. The outer capsule contains an oxygen buffer, such as magnetite–hematite or nickel–nickel oxide, together with pure water. This fixes f_{H_2} in the charge by hydrogen diffusion through the platinum membrane. The chlorine fugacity in the charge is fixed by the equilibrium between silver and silver chloride:

$$2AgCl \rightleftharpoons 2Ag + Cl_2$$

and the HCl fugacity by the externally imposed f_{H_2}:

$$Cl_2 + H_2 \rightleftharpoons 2HCl$$

The final constraint is that total pressure is everywhere the same so that we have:

$$P_{total} = P_{H_2O} + P_{H_2} + P_{HCl} + P_{O_2} + P_{Cl_2}$$

where P_i is the partial pressure of gas i. In practice, for most of the experiments, H_2O is by far the dominant species and the partial pressure of water is very close to the total pressure. The molality of HCl in these buffered high $P-T$ solutions was calibrated as a function of P, T, and f_{H_2}

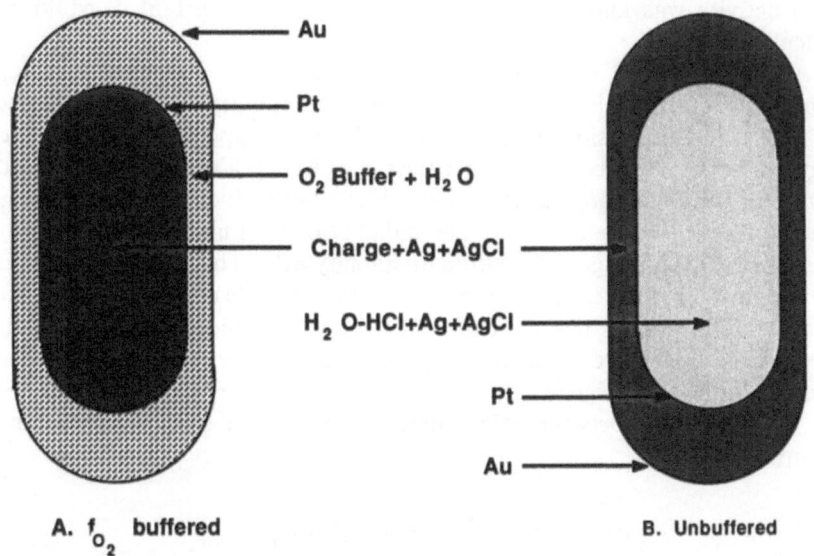

Figure 5.10 Cross-section views of capsule arrangements using silver–silver chloride sensors. (a) f_{O_2} buffered system. (b) System without f_{O_2} buffer. From Frantz and Eugster (1973).

by Frantz & Eugster (1973), Chou & Frantz (1977), and Frantz & Popp (1979). [See Chou (1987) for a discussion of the technique.] These calibrations were performed with a charge consisting solely of Ag + AgCl in a rapid-quench cold-seal pressure vessel. On quenching, the capsules were punctured, the solution extracted with a micropipette, and total chloride in solution determined with an ion-specific electrode. In this approach, the total chloride in the quenched solution is equal to total HCl present at the P and T of the experiment.

The calibrations of the Ag + AgCl buffer described above may be used in the manner described by Frantz & Popp (1979) to determine the equilibrium constants for silicate dissolution reactions such as Reaction (5.9). Frantz & Popp actually studied a different but analogous reaction:

$$\tfrac{1}{3} Mg_3Si_4O_{10}(OH)_2 + 2HCl \rightleftharpoons MgCl_2 + \tfrac{4}{3}SiO_2 + \tfrac{4}{3}H_2O \qquad (5.10)$$

$$\text{talc} \qquad\qquad \text{fluid} \qquad \text{fluid} \qquad \text{quartz} \qquad \text{fluid}$$

Since the molality of HCl was buffered, the equilibrium constant for Reaction 5.10 was obtained by piercing the inner capsule and determining the magnesium contents of the resultant solutions. In practice this was done by atomic absorption spectrometry. Provided all of the Mg was present as $MgCl_2$ at the P and T of the experiment, the Mg concentration gave $MgCl_2$;

HCl activity was known, and thus the equilibrium constant K could be calculated.

A variant of the method (Fig. 5.10b) is aimed at reducing the HCl concentrations which are often outside the stability range of silicate minerals. In this version the charge is in the outer capsule and there is no oxygen buffer. The combination of $Ag + AgCl + H_2O$ in both inner and outer capsules fixes m_{HCl} at the same value in both, since the Pt membrane keeps f_{H_2} everywhere constant. The molality of HCl is then controlled by the mineral assemblage to be within its stability field. The molality of HCl is determined after the experiment by measuring chloride concentration in the inner capsule which has an H_2O–HCl fluid. The molality of $MgCl_2$ is determined as before by measuring Mg concentration in the solution in contact with the charge.

At equilibrium between talc, quartz, and fluid at some pressure and temperature

$$G_{MgCl_2}^{fluid} + \frac{4G_{qtz}}{3} + \frac{4G_{H_2O}}{3} = \frac{1G_{talc}}{3} + 2G_{HCl}^{fluid} \qquad (5.11)$$

Since talc and quartz are pure the free energies of these phases at the P and T of interest can be substituted directly into Equation 5.11. The free energy of H_2O may be calculated by taking the value for pure H_2O and making a small Raoult's Law correction for the contents of $MgCl_2$ and HCl:

$$G_{H_2O} = G_{H_2O}^0 + RT \ln X_{H_2O}^{fl}$$

These substitutions enable calculation of the free energy difference $(G_{MgCl_2}^{fl} - 2G_{HCl}^{fl})$ at high P and T. Frantz & Popp took these values and converted them to a hypothetical 1 molal standard state assuming unit activity coefficients as follows:

$$G_{MgCl_2}^{fl} = G_{MgCl_2,fl}^0 + RT \ln m_{MgCl_2}$$

They then derived values of the standard state free energy difference which at 1 kbar is:

$$G_{MgCl_2,fl}^0 - 2G_{HCl,fl}^0 = 619.1 \ (\pm 11.5) + 0.3192 \ (\pm 0.015)T \ \ \text{kJ. (5.12)}$$

In Equation 5.12 the free energy of formation refers to formation of a hypothetical 1 molal solution from the elements at the P and T of interest.

Having turned the mineral–fluid equilibria into a free energy difference for the solution species, thermodynamic data for other solids may be used

Table 5.2 Solubility constants of some minerals (from Frantz *et al.* 1981).

Chain silicates	Log $K = A + B/T(K)$	Pressure (kbar)	A	B
Wollastonite	$CaSiO_3 + 2HCl^0 + H_2O \rightleftharpoons CaCl_2^0 + H_4SiO_4$	1.0	-7.547	8,287.4
		2.0	-5.773	7,754.1
Ca–Al pyroxene	$CaAl_2SiO_6 + 2HCl^0 + H_2O \rightleftharpoons CaCl_2^0$ $+ Al_2O_3 + H_4SiO_4$	1.0	-8.252	8,920.4
		2.0	-6.477	8,375.9
Jadeite	$NaAlSi_2O_6 + HCl^0 + 3.5H_2O \rightleftharpoons NaCl^0$ $+ 0.5Al_2O_3 + 2H_4SiO_4$	1.0	-3.181	3,039.5
		2.0	-1.116	2,178.6
Enstatite	$MgSiO_3 + 2HCl^0 + H_2O$ $\rightleftharpoons MgCl_2^0 + H_4SiO_4$	1.0	-13.081	10,844.0
		2.0	-8.863	8,562.0
Ferrosilite	$FeSiO_3 + 2HCl^0 + H_2O \rightleftharpoons FeCl_2^0 + H_4SiO_4$	1.0	-8.579	8,619.3
		2.0	-4.853	6,278.9
Diopside	$CaMgSi_2O_6 + 4HCl^0 + 2H_2O \rightleftharpoons CaCl_2^0$ $+ MgCl_2^0 + 2H_4SiO_4$	1.0	-19.665	17,347.3
		2.0	-13.929	14,689.7
Hedenbergite	$CaFeSi_2O_6 + 4HCl^0 + 2H_2O \rightleftharpoons CaCl_2^0$ $+ FeCl_2^0 + 2H_4SiO_4$	1.0	-15.978	16,344.2
		2.0	-10.480	13,447.1
Anthophyllite	$Mg_7Si_8O_{22}(OH)_2 + 14HCl^0 + 8H_2O$ $\rightleftharpoons 7MgCl_2^0 + 8H_4SiO_4$	1.0	-88.116	71,232.6
		2.0	-57.865	54,853.7
Tremolite	$Ca_2Mg_5Si_8O_{22}(OH)_2 + 14HCl^0 + 8H_2O$ $\rightleftharpoons 2CaCl_2^0 + 5MgCl_2^0 + 8H_4SiO_4$	1.0	-72.469	60,074.7
		2.0	-48.681	48.291.9
Pargasite	$NaCa_2Mg_4Al_3Si_6O_{22}(OH)_2 + 13HCl^0$ $+ 4.5H_2O \rightleftharpoons NaCl^0 + 2CaCl_2^0 + 4MgCl_2^0$ $+ 1.5Al_2O_3 + 6H_4SiO_4$	1.0	-68.760	60,386.0
		2.0	-53.471	53,613.0

to calculate any equilibrium of interest. Frantz *et al.* (1981) present values of equilibrium constants for a wide range of mineral–solution reactions involving HCl, $MgCl_2$, $FeCl_2$, $CaCl_2$, NaCl and KCl. A few of the data are presented in Table 5.2 as equations relating log K to temperature. In Table 5.2 standard states for solids are the pure phase at the P and T of interest. For solutes ($MgCl_2$, etc.), a hypothetical 1 molal solution at P and T is used, and for H_2O the pure phase at the P and T of interest is the standard state.

The importance of these solution experiments and their derived log K values is that they provide the basis for calculation of the concentrations of species in equilibrium with solids at high pressure and temperature, provided an estimate of total chloride in solution may be made. Values of the latter obtained by Poty *et al.* (1974) are up to about 5% NaCl equivalent in fluid inclusions. The major problem is that the log K values do not work in the region between about 200 and 450°C where both molecular $MgCl_2$,

NaCl, etc. and ionic Na^+, Mg^{2+}, Cl^- species are present. In order to make calculations of mass transport in this temperature range it is necessary to have precise data on the dissociation constants of the fluid species. This requires *in situ* measurement of, for example, electrical conductance of the fluid in the *P–T* range of concern. Experiments of this type will be discussed in the next chapter.

6 Just fluids

6.1 Objectives

Fluids are among the most effective agents for the transport of material in the Earth's crust and, possibly, the upper mantle. Aqueous, hydrothermal fluids are responsible for the formation of most ore deposits and for much of the control of sea water chemistry. In addition, fluids play a major role in many metamorphic reactions. Prediction of the behavior of radioactive and hazardous waste in hydrothermal environments requires understanding of fluid chemistry. Fluids, as used in the present context, are most often H_2O-rich, and are either liquids or, at temperatures above about $400°C$, supercritical fluids. These fluids have minor to major amounts of dissolved NaCl and may contain significant amounts of CO_2 or CH_4. In order to understand processes involving fluids, we need to know the thermodynamic properties of their major and minor constituents, the solubilities of a wide range of minerals, and how those solubilities change with P, T, pH, and f_{O_2}. We must also know their transport properties, such as viscosity and thermal conductivity. A good understanding of fluid properties requires identification of the molecular grouping, or speciation, of elements in fluids and how that speciation changes in response to P, T, and the internal stucture of the fluid as a whole. In this chapter we begin by describing experiments to measure properties of the major molecular constituents, especially H_2O, and then examine some of the methods of determining the behavior of minor species.

6.2 Volumetric measurements: $P-V-T$

$P-V-T$ experiments consist of the simultaneous measurement of the pressure, volume, and temperature of a phase. Theoretically, knowledge of the $P-V-T$ properties of a phase as a function of P and T, together with knowledge of the heat capacity of the phase at 1 atm, completely describe the thermodynamic properties of the phase. In practice, we can use $P-V-T$ measurements to describe the thermodynamic properties of single component fluids, such as pure H_2O or pure CO_2, but in many cases the extension even to binary systems is difficult, and multicomponent systems are out of the question.

A powerful result of $P-V-T$ experiments is the direct measurement of the volume of a phase and hence its density (density = 1/specific volume).

91

Density appears to be the most important variable affecting the viscosities, dielectric constants, and heat capacities of fluids. The density itself must be known for fluid-dynamic modeling of, for example, hydrothermal systems.

The measurement of P, V, and T for a fluid is straightforward in principle. Two approaches have been used. In one, the volume of the experimental system is fixed (by using a rigid sample chamber), and P is measured as a function of T. In the second approach, the sample chamber may expand or contract, and its volume is measured as a function of P and T. The fixed volume type of apparatus is very simple in construction, and experiments made using it can be very precise and accurate. Unfortunately, such systems can be used only in the elastic range of vessel materials, and all suitable materials become soft and deform plastically at high temperatures. Consequently, under high P–T conditions, the more complicated, variable volume type of system must be used.

6.2.1   P–V–T *of single-component fluids:* H_2O

Because H_2O is the most common major component of natural fluids, its $P-V-T$ properties were the first to be studied over a wide range of geologic $P-T$ conditions. The experiments by Burnham *et al.* (1969) illustrate the use of a variable volume cell which allowed direct volumetric determinations to 900°C and 8,500 bar. A schematic diagram of the $P-V-T$ apparatus is shown in Figure 6.1. The entire volumetric apparatus is placed in an internally heated, argon-medium vessel. This apparatus is a considerable modification of the usual variable volume type of system. In fact, the volume of the high T cell remains nearly constant, while the volume of the

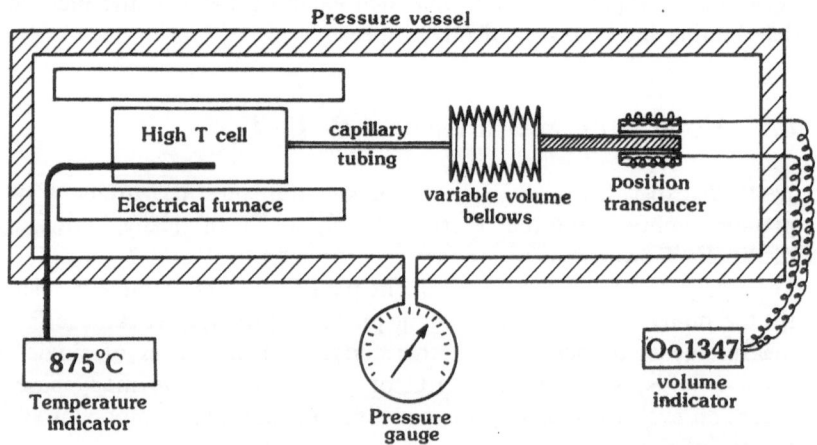

Figure 6.1   Schematic diagram of the $P-V-T$ apparatus used by Burnham *et al.* (1969) to measure H_2O.

room T bellows changes in response to changes in P and T. The reason that this system works at temperatures in the plastic region of the cell is that the pressure on the inside and outside of the cell is identical, so no force exists to cause deformation of the cell. It was necessary to make corrections to the volume of the high T cell for the compressibility and thermal expansion.

One of the most useful thermodynamic properties obtainable from the $P-V-T$ data is the change in free energy as a function of P at constant T:

$$\Delta G_{H_2O} = G_{P_2} - G_{P_1} = \int_{P_1}^{P_2} V_{H_2O} \partial P \qquad (6.1)$$

and because fugacity is related to ΔG,

$$f_{H_2O} = \exp\left[\int_{P_1}^{P_2} V_{H_2O} \partial P\right]_T \qquad (6.2)$$

Equation 6.2 was used to calculate the fugacity tables of Burnham et $al.$ (1969a). Because fugacity is calculated by integrating V over a pressure range, small errors in V tend to be canceled out, and consequently the calculated fugacities for single component fluids have high precision and accuracy.

6.2.2 P–V–T of two-component fluids

Seward & Franck (1981) measured the $P-V-T$ properties of mixtures of H_2O and H_2 at T up to $440°C$ and 2,500 bar. These $P-T$ conditions are within the elastic limit of the pressure vessel material they were using, so that the fixed volume method could be used. A schematic diagram of their system is shown in Figure 6.2. The system consists of an externally heated vessel (Section 3.4) which has an accurately known volume, a screw press for injecting known volumes of H_2O into the vessel, and a hydrogen pump to pressurize the vessel with H_2. The pressure vessel was also equipped with sapphire windows so that the transition from the one–phase (supercritical fluid) to the two-phase region (aqueous liquid and H_2-rich gas) could be observed visually.

An experiment consisted of adding known amounts of H_2 and H_2O to the vessel, closing the vessel off and heating to about $450°C$. The pressure attained was determined by the amounts of H_2 and H_2O loaded into the vessel at the beginning of the run. After the system had stabilized, the T was gradually lowered and T and P continuously recorded. In order to determine the system, many combinations of $H_2 : H_2O$ ratio and initial amount were necessary.

The first-order thermodynamic question to be asked about fluid mixtures is, do they mix ideally? If they do mix ideally, the volume of the mixture is a

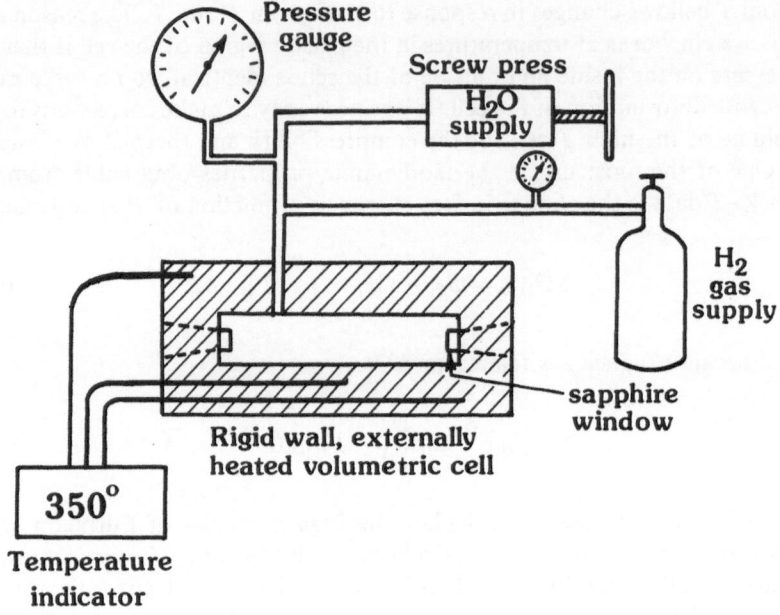

Figure 6.2 Schematic diagram of the $P-V-T$ system used by Seward and Franck (1981) to measure the H_2-H_2O system.

linear combination of the volumes of the end-members, so that the total excess volume of the mixture:

$$V_T^E = V_{measured} - X_{H2} \cdot V_{H_2} - X_{H_2O} \cdot V_{H_2O}$$

is equal to zero. Figure 6.3 is a plot of excess volumes from Seward & Franck (1981). There is definitely a deviation from ideality at 400 bar, but at 2,000 bar it is not clear whether positive and negative deviations are present or if the mixtures are ideal within the precision of the volume measurements ($\pm 0.1\%$).

The excess volume plot can be used to extract activities of H_2 or H_2O in the mixture. Choosing the $P-T$ standard state for $a_{H_2O} (= f_{H_2O}/f_{H_2O}^0$, where $f_{H_2O}^0$ is the fugacity for pure H_2O at the P and T of interest), the activity is given by

$$RT \ln a_{H_2O} = \int_{P_1}^{P_2} V_{H_2O}^E \partial P \tag{6.3}$$

where $V_{H_2O}^E$ is the partial molar excess volume of H_2O, which is calculated

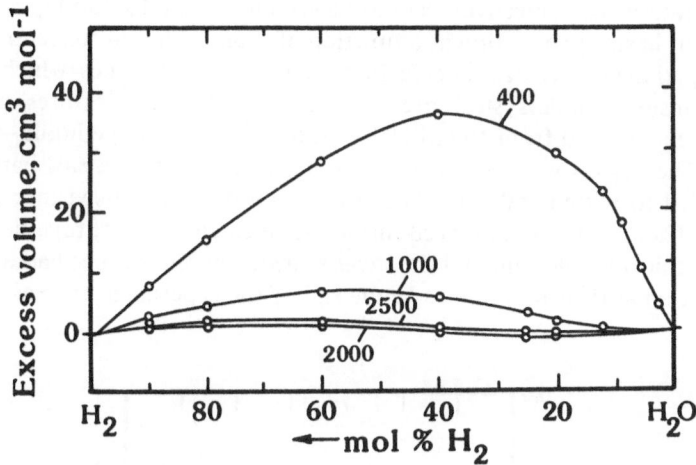

Figure 6.3 Excess total volumes for H_2–H_2O solutions at $400°C$. Adapted from Seward and Franck (1981).

from

$$V^E_{H_2O} = (V^E_T)x_{H_2O} + \left(\frac{\partial V^E_T}{\partial_{H_2O}}\right)_{x_{H_2O}} \cdot (1 - X_{H_2O})$$ (6.4)

where V^E_T, the total excess volume, and the derivative $(\partial V^E_T/\partial_{H_2O})$ are taken from plots such as Figure 6.3 at the values of X_{H_2O} in question. Note that the partial molar excess volumes must be integrated from 1 bar to the pressure in question. The difficulty in obtaining activities in this manner is that data must be available over the pressure range of interest and, equally important, the precision of the volume determinations must be very high because of the need to take derivatives of the total excess volume. These difficulties have greatly limited the use of $P-V-T$ measurements to obtain activity–composition relations in fluid mixtures.

6.3 Speciation from solubility

$P-V-T$ measurements are powerful tools for extracting properties of the major components of fluids, but are of little use in determining the nature of components in minor or trace amounts. In the next three sections we discuss methods to identify the species present in aqueous solution. We begin with an example of a simple but often powerful method, the measurement of solubility.

Seward (1976) measured the solubility of silver chloride in aqueous sodium chloride solutions as a function of temperature and chloride concentration. His objectives were to determine the total solubility and the variations in silver speciation as a function of T and chloride concentration. The experiments were conducted in sealed silica glass tubes which were placed inside stainless steel pressure vessels. The temperatures of the experiments ranged from 100 to 350°C at pressures along the liquid–vapor curve. The vessels were placed in a thermostated oven or a salt bath and controlled to within ±0.5°C. The silver chloride was pressed into glassy pellets. These pellets were placed inside the silica glass tubes, together with sodium chloride solution, and the tubes sealed. The silica tubes had a slight constriction at their mid-point. At the end of the experiment the vessel was

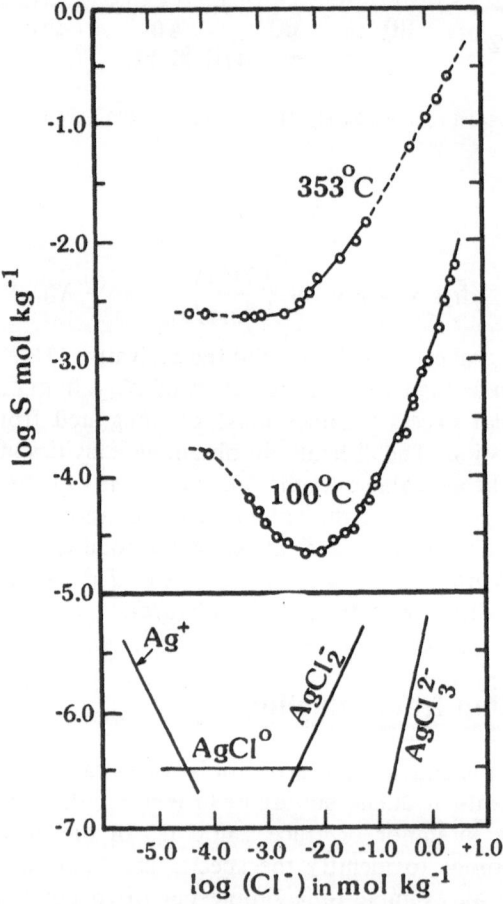

Figure 6.4 Log of the total silver solubility *vs.* log of chloride content of the solution. Experimental data for 100 and 353°C. The lower portion of the diagram shows theoretical slopes for various silver species. Adapted from Seward (1976).

turned upside down, the silver chloride pellet was caught in the constriction, and the solution drained to the bottom end of the tube where it could not react with the pellet.

The slope of a line on a plot of log S (total solubility) against log a_{Cl^-} is expressed as

$$\frac{d \log S}{d \log a_{Cl^-}} = n - 1 \qquad (6.5)$$

where n is the number of chloride ions coordinated to the central silver ion. If the activity coefficient for Cl^- remains constant with changes in Cl^- concentration, then chloride activity (a_{Cl^-}) is directly proportional to Cl^- molality, and Equation 6.5 expresses the slope of the experimental data shown in Figure 6.4. The heavy lines shown in Figure 6.4 represent slopes for several chloride species. By comparing the theoretical slopes with the experimental data, the silver speciation and the change in speciation with T can be deduced.

6.4 Speciation from spectroscopy

It is apparent from the discussions in Chapter 5 and Section 6.1 that direct measurements of speciation in high temperature solutions are important to the interpretation of solubility data and the understanding of mass transfer in geothermal and metamorphic systems. One method of measuring speciation involves the use of *in situ* spectroscopic techniques. The complexes formed at high P and T have characteristic absorption bands due to molecular vibration and rotation and to ligand–metal charge transfer. If these bands can be measured and quantitatively deconvoluted, they provide an excellent means of determining the relative proportions of different species at high P and T.

A simple apparatus for spectrophotometric measurements up to about $300°C$ is shown in Figure 6.5 (Giggenbach 1971; also see Buback *et al.* 1987). It consists of a split steel block with a gold liner forming the sample chamber. Pressure is maintained by the upper sealing nipple and two windows which are held in place by steel nipples with holes through their middles. The windows are made of thick silica glass, sapphire, or some other material transparent to the working wavelength range. Heating is by four nichrome elements passed through holes in the steel block. The whole apparatus is placed inside a spectrophotometer and radiation shone through the transparent window. As might be anticipated, the major limitation of this type of cell is the stength of the window. More sophisticated designs for use in the supercritical region generally employ sapphire windows (Buback 1981).

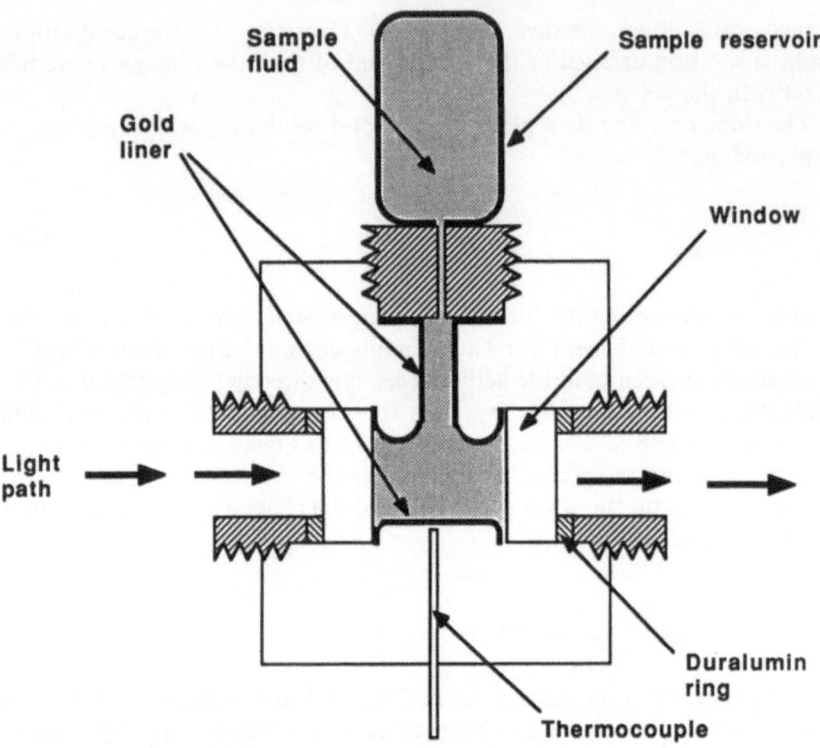

Figure 6.5 Cross-sectional view of a moderate $P-T$ spectroscopic cell (Modified from Giggenbach, 1971).

Seward (1984) used the apparatus of Figure 6.5 to investigate the ultraviolet spectra of lead (II) complexes in sodium chloride solution. The ligand to metal charge transfer bands in such solutions vary from complex to complex. For example, at $25°C$, the Pb^{2+} absorption band is centered at a wavelength of 209 nm, $PbCl^+$ at 226 nm, $PbCl_4^-$ at 268 nm, and so on (Seward 1984). Seward determined the absorption spectra of solutions containing very small amounts of lead, about 1×10^{-4} molal and with chloride concentrations from 1.2×10^{-3} to 3.223 molal. The reason for using a range of chloride concentrations is that, depending on the dissociation constants for the lead chloride complexes, increasing Cl^- increases the amounts of the higher complexes through the equilibria:

$$Pb^{2+} + Cl^- \rightleftharpoons PbCl^+$$
$$PbCl^+ + Cl^- \rightleftharpoons PbCl_2^0$$
$$PbCl_2^0 + Cl^- \rightleftharpoons PbCl_3^-$$
$$PbCl_3^- + Cl^- \rightleftharpoons PbCl_4^{2-}$$

98

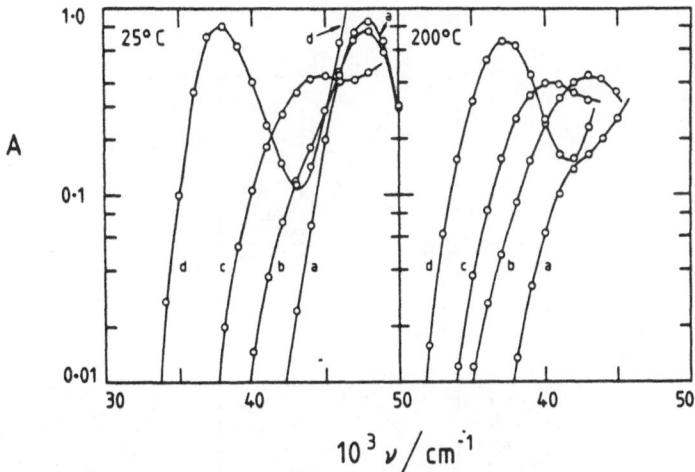

Figure 6.6 Spectra showing absorbances for solutions containing 1×10^{-4} M lead and having chloride concentrations of (a) 0.0012 M (b) 0.010 M (c) 0.10 M (d) 3.223 M. From Seward (1984).

Hence shifts in the absorption spectra with chloride content provide information on where the absorption band is for each complex.

Figure 6.6 shows spectra for temperatures of 25 and 200°C at the chloride molalities used by Seward. Distinct effects of temperature and concentration can clearly be seen, although it is not intuitively obvious what these effects correspond to in terms of complexing. However, with spectra obtained at a range of Cl⁻ concentrations, and bearing in mind that absorption energy tends to shift regularly with increasing numbers of Cl⁻ attached to Pb, deconvolution of the spectra can be attempted. The results of this process are shown in Figure 6.7, which is a plot of percentages of the different complexes as a function of chloride content at temperatures from 25 to 300°C. It may be seen that the uncomplexed Pb^{2+} ion is very important at 25°C and low chloride content, but that its concentration diminishes as temperature increases. This is consistent with the observation that increased temperature causes the formation of complexes from ionized species. Another interesting observation is that with increasing temperature the species with most attached chloride, $PbCl_4^{2-}$ also diminishes in importance, being supplanted by $PbCl_2^0$ and $PbCl_3^-$. This demonstrates two effects, a decrease in Cl⁻ activity due to complexing with Na^+ and a tendency to stabilize complexes with near neutral charge as the dielectric constant of water decreases.

Raman and infrared (IR) spectra may provide similar or complementary information to that from ultraviolet spectra. The energies involved are less than one-tenth of those due to ligand to metal charge transfer and correspond to vibrations of, for example, Pb—Cl bonds or O—H bonds.

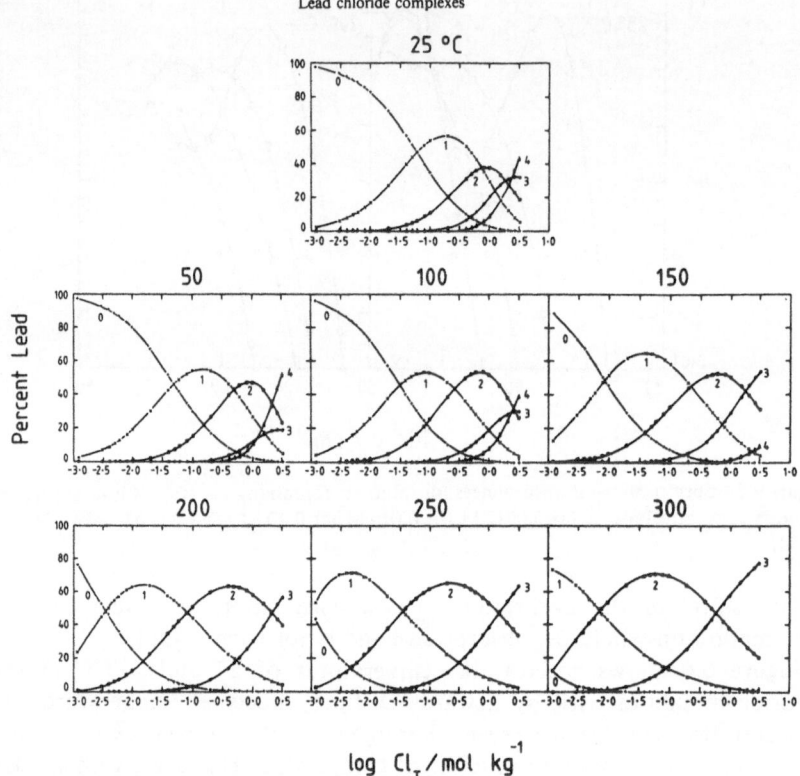

Figure 6.7 Relative abundances of lead chloride complexes as a function of Cl_T from 25 to $300°C$. The curve for each complex is labeled according to number of chloride ions, i.e., $2 = PbCl_2$. From Seward (1984).

As such they may be used to obtain spectra corresponding to complexes like those we have already discussed, or they may be used to look at the major components of the fluid such as H_2O or CO_2. In the case of IR spectra there are severe restrictions on window materials because most available windows are only transparent to IR in limited wavelength ranges (Buback 1981). High-temperature Raman spectra can, on the other hand, be taken with sapphire windows because all frequencies involved are in the visible region where such windows are transparent.

Figure 6.8 is taken from Franck (1974) and shows IR absorption due to the oxygen—deuterium vibration in fluids containing H_2O diluted with HDO. Deuterated water was used because the O—D spectral band is free from interference in the frequency range studied. However, any compositional or temperature effects on the O—D band may be regarded as analogous to those on O—H. The figure compares the effect of diluting H_2O with xenon

100

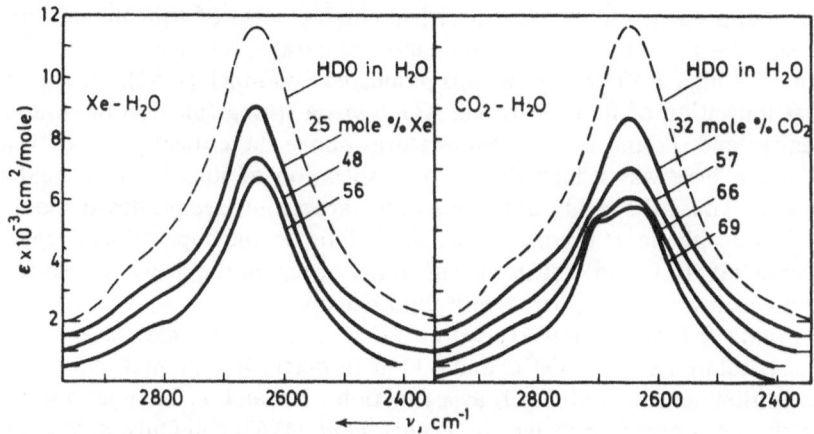

Figure 6.8 The effects of xenon and CO_2 on the O—D vibration in deuterated water at $400°C$. From Franck (1984).

with that obtained when CO_2 is the diluant. Xenon is of similar size to CO_2, but being a non-polar inert fluid is not expected to interact as strongly with H_2O as the latter. It may indeed be seen from Figure 6.8 that the effect of diluting H_2O at $400°C$ with Xe is simply to reduce the intensity of the O—D absorption band. No strong chemical interactions appear to occur. The influence of CO_2 on the band is, as anticipated, much more marked. A shoulder at $2,700$ cm^{-1} appears at 57 mol% CO_2 and this changes into a separate band at 69 mol%. The high energy band is possibly due to the formation of a distinct form of carbonic acid complex, H_2CO_3. (Note that the scale on Figure 6.8, cm^{-1}, "wavenumber," is the reciprocal of wavelength. It is often used because, unlike wavelength, the wavenumber scale is linear in energy).

Similar Raman and IR studies of pure H_2O or aqueous electrolytes reveal information about other structural changes in the solvent (e.g., Franck 1973). The O—D stretching frequency moves to higher energy with increasing temperature (at constant density) due to decreasing strength of hydrogen bonding. Addition of a salt such as KI also produces an increase in stretching frequency at $25°C$ due to the fact that dissolved ions also tend to break up the structure of water. At higher temperatures the structure-breaking tendency of the solute is less marked (Franck 1973).

6.5 Speciation from electrical conductivity

It was noted in Chapter 5 (Section 5.4.2) that determining ionization constants (and hence speciation) from solubility measurements in multi-component systems may sometimes yield erroneous results. A much more direct approach is to measure electrical conductances of aqueous solutions at high $P-T$ conditions. Development of apparatus for measuring conductances at high $P-T$ conditions was pioneered by Franck (1956). He showed that ionization of KCl, HCl, and KOH was a strong function of solution density and T, and that at temperatures above the critical point of H_2O (374°C) there was a high degree of association of ions to form neutral species. Thus salts, acids, and bases which are strong electrolytes at room T and atmospheric P become weak electrolytes in the supercritical region. Franck (1956) found that in the supercritical region the degree of association was inversely correlated with fluid density.

Frantz & Marshall (1982) used electrical conductance measurements of $CaCl_2$ solutions up to 600°C and 4 kbar to determine the first and second ionization constants of $CaCl_2$ as a function of P and T. They used a high pressure cell based on designs by Franck *et al.* (1962) and Quist & Marshall (1968). The cell, shown schematically in Figure 6.9, consists of a platinum–iridium lined cold-seal vessel (Section 3.4) with a small Pt–Ir cylinder centered in the vessel. The small cylinder forms one electrode and the vessel liner the other. The electrodes are coated with platinum black to minimize polarization effects. For concentric cylinders, specific conductivity (σ) is related to the measured resistance (R) between the electrodes by

$$\sigma = \frac{1}{R} \cdot \frac{1}{2\pi L} \cdot \ln \frac{r_2}{r_1} \qquad (6.6)$$

where L is the length of the small electrode cylinder and r_2 and r_1 the inside diameter of the vessel and the outside diameter of the small electrode. Accurate measurement of σ required determining cell constants using solutions of known conductivities. Sample solutions were made with 0.001, 0.0025, and 0.005 molal (mol/kg) concentrations of $CaCl_2$. In order to calculate ionization constants, the specific conductance values must be

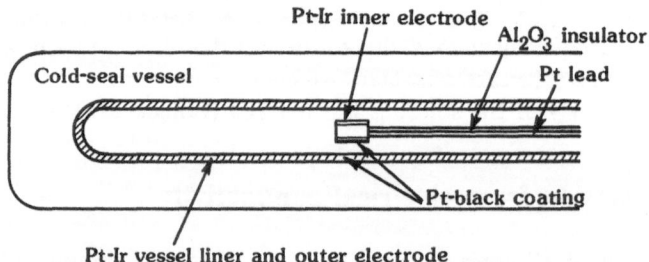

Figure 6.9 Schematic diagram of the electrical conductance cell used by Frantz and Marshall (1982) to measure aqueous $CaCl_2$ solutions.

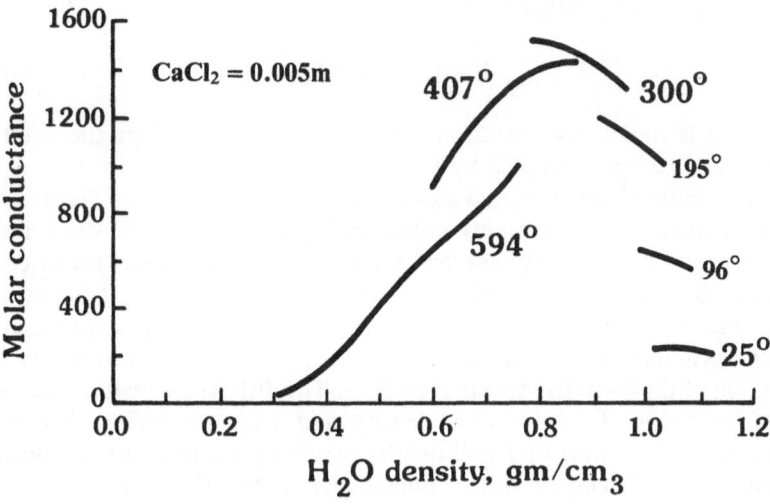

Figure 6.10 Molar conductances ($cm^2\,ohm^{-1}\,mol^{-1}$) of $CaCl_2$ as a function of density at constant temperature (°C). From Frantz and Marshall (1982).

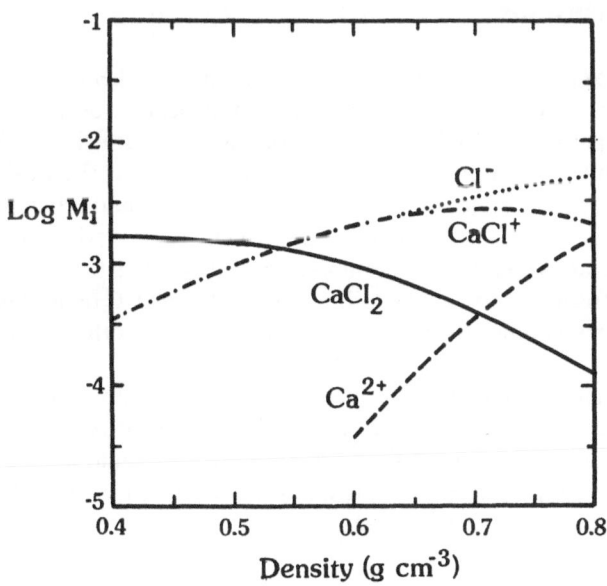

Figure 6.11 The logarithm of the molarity of $CaCl_2$, $CaCl^-$, Ca^{2+}, and Cl^- as a function of density for a 0.005 molal $CaCl_2$ solution at 500°C. From Frantz and Marshall (1982).

103

converted to molar conductances as follows

$$\lambda = 1{,}000 \; \sigma / m d_{H_2O} \tag{6.7}$$

where m is molal concentration and d_{H_2O} is the density of the solution (which can be approximated by the density of pure H_2O).

An example of the measured molar conductances is shown in Figure 6.10. Molar conductance is strongly influenced by the viscosity of the solution and the concentrations of charged species. At constant density the viscosity of H_2O decreases rapidly with increasing T until the critical T is reached. At higher temperatures viscosity is independent of T. The concentration of charged species is determined by the degree of ionization, which is a function of the dielectric constant of H_2O. The dielectric constant decreases with increasing T at constant density and with decreasing density at constant T. The effect of T and density on viscosity and dielectric constant result in the behavior of molar conductance seen in Figure 6.10.

Frantz & Marshall (1982) show how molar conductances are used to calculate the degree of dissociation of $CaCl_2$ into $CaCl^+$ and Cl^-, and of $CaCl^+$ into Ca^{2+} and Cl^-. These results can then be used to generate a plot such as Figure 6.11 in which the concentration of species is shown as a function of fluid density.

6.6   Conclusions

Experimental techniques available to investigate the chemistry of fluids at moderate P–T conditions are now quite sophisticated. Some of the other types of experiments which we have not discussed are flow calorimetry, polarography, and isopiestic measurements. In spite of the availability of these techniques, few measurements have been made, and much more work will have to be done in order to accurately calculate speciation in hydrothermal systems at even moderate temperatures. The transition from strongly ionized to strongly associated species in the 300 to 600°C temperature range makes understanding of processes in this range both interesting and complicated.

Our understanding of fluid properties above a few thousand bar is much more limited and should prove to be a fruitful area for research. Certainly many of the techniques used at moderate pressures could be, and in some cases have been, adapted to piston-cylinder pressure vessels, allowing studies in the 10 to 40 kbar range.

7 Igneous experiments on melts and crystals

7.1 Introduction

7.1.1 Experimental objectives

The first igneous experiments were done just to prove that lavas were once molten (this obviously would not have been necessary if the experimenter had lived on Hawaii or Iceland). Tuttle & Bowen (1958) had the same objective when they proved that granites could form from magmas containing significant amounts of H_2O. The objectives of modern experimental igneous petrologists include the following:

(a) To determine the processes occurring during magma evolution such as crystal fractionation, magma mixing, etc. This includes determining the P, T, and intensive variable conditions under which the processes occur, and the relationships of the processes to tectonic models.
(b) To use magmas as windows to their source regions at depth. If one can "see through" the evolutionary events in the history of magmas, they can be used to estimate the P and T of their sources in the lower crust or upper mantle. Knowing the $P-T$ conditions at the source constrains its thermal history.
(c) To use magma composition to constrain the chemical make-up of the source regions.

7.1.2 Approaches used

Historically, igneous experiments have been of two types: studies of compositionally simple systems which are of general applicability, and studies of individual natural rock samples. Much information has been gained from each type, but simple systems are not adequate to reach a full understanding of natural magmas, while studies on a single rock are difficult to generalize to the range of magma compositions encountered in nature.

The preferred approach is to construct thermodynamic models for silicate melts and crystals and then to design experiments specifically to test the models. The experiments can be done on simplified systems designed to test one, or a few, parameters at a time. Most of the examples given here were not preceded by a thermodynamic model, but we have chosen them to illustrate the best experimental techniques.

7.1.3 Experimental conditions

Different magma types vary with respect to the volatile content and pressure range important in their origin and evolution. For example, basaltic rocks collected from the lunar surface and Earth's ocean floors appear to have formed under essentially volatile-free conditions. For this reason, their low P fractionation can be modeled by experiments at atmospheric pressure. On the other hand, an H_2O content of 3 to 5 wt% is inferred for most andesitic to granitic magmas, and the hydrous minerals biotite and hornblende are important in their evolution. Consequently, experiments bearing on their evolution require pressures appropriate to the continental crust (up to 10 kbar), and control of volatile activities. Basic alkaline magmas have their origins in the upper mantle, and both CO_2 and H_2O may have been important in their formation. Thus, experiments under controlled volatile activities at pressures of at least 30 kbar are required.

Examples of experiments will be given in order of pressure range, beginning with atmospheric P and progressing to crustal and then upper mantle pressures.

7.2 Atmospheric pressure experiments

7.2.1 The simple system diopside–albite–anorthite

We begin by examining one of the classic systems and describe how it has been extended by recent thermodynamic measurements. N. L. Bowen, the father of experimental petrology, chose this system because it contains the two major minerals which make up basaltic rocks – clinopyroxene and plagioclase – while being compositionally simple. It also does not contain iron, so oxygen fugacity has no effect on the system and the experiments can be done in air.

The experiments of Bowen (1915) were done in quench furnaces (Section 3.2) as were all later equilibrium experiments in this system. The starting compositions were glasses prepared by several crushing and grinding cycles and then partly or completely crystallized to mixtures of plagioclase and diopsidic pyroxene. All of the compositions used by Bowen contain Na, and this element is volatile enough at high T that significant loss of Na would occur if the molten samples were exposed to the air. To prevent Na loss, he enclosed the samples in Pt foil as later described by Schairer (1951). For each composition, Bowen found the T at which the last crystal disappeared by doing a series of experiments at increasing temperature. After each experiment the sample was crushed in an agate mortar and pestle and examined as a grain mount in transmitted light. Even small amounts of crystals are easily seen with this technique, so that the liquidus T can be precisely determined. Because the starting materials contained both plagioclase and clinopyroxene these are melting experiments, and any disequilibrium will result in the observed liquidus T being higher than the true value. There is no evidence of significant inaccuracy due to this effect except in the most albite-rich portion of the system. From his experimental data Bowen was able elegantly to explain crystallization behavior of systems containing solid solutions. For example, as can be seen in Figure 7.1, Bowen's results, when applied to rocks, correctly predict the coexistence of clinopyroxene with a plagioclase which becomes increasingly sodic as fractional crystallisation proceeds. His treatment of this system is a classic example of the power of well executed experiments in simple systems.

Before the development of the electron microprobe, there was no means available for chemically analyzing the complex liquid and crystals in experimental run products because they were too fine-grained to separate

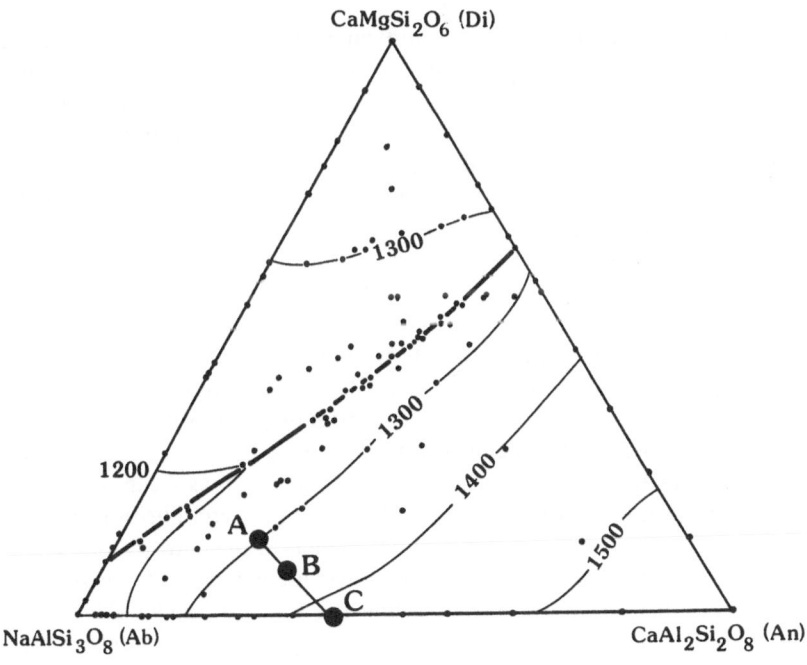

Figure 7.1 The Di–Ab–An system showing the compositions used in phase equilibrium experiments. The liquidus isotherms are calculated using the thermodynamic model discussed in the text. Di = diopside, Ab = albite, An = anorthite. Based on Weill *et al.* (1980).

for conventional analysis. Thus Bowen was forced to do a number of experiments at different T for each bulk composition he studied. With the electron microprobe, coexisting liquid and crystals in a single run can be analyzed, and the liquidus T and composition may be determined from a single well planned experiment (providing the position of the liquidus is roughly known beforehand). The result is illustrated in Figure 7.1 where the analyzed liquid and plagioclase compositions are plotted together with the bulk composition. The liquid composition (point A) gives the liquidus position, and the plagioclase composition (point C) is that of the last crystal before composition A is completely molten at the run T, $1,400°C$.

Crystal/liquid equilibria in magmatic systems may, in principle, be calculated from thermodynamic models. Because the diopside–albite–anorthite (di–ab–an) system is so well studied and contains two important magmatic minerals, it was one of the first systems used to test the thermodynamic approach (Weill *et al.* 1980). The approach is based on the equality of the chemical potentials of components in crystals and melt. Using standard thermodynamic relations, they derived the following equality (we use diopside for an example):

$$RT \ln a_{di}^{Cpx} = \Delta_f H_{di} - T \Delta_f S_{di} + \Delta_m h_{di}^{liq} - T \Delta_m S_{di}^{liq}$$

The same equality holds for ab and an components in plagioclase. The activity–composition relations of components in crystals have been discussed in Chapter 4, and have been at least approximately determined. The four quantities on the right side must be determined in order to solve for T. The heat of fusion, $\Delta_f H_{di}$ and entropy of fusion $\Delta_f S_{di}$ can be determined calorimetrically or estimated from the melting curve of the pure component as a function of P and T using molar volumes and high T heat capacities of melt and crystal. All of these values have now been measured for di, ab, and an. Beginning with Weill *et al.* (1980), Navrotsky and coworkers have systematically measured the partial molar heats of the components ($\Delta_m h_{di}^{liq}$) in the di–ab–an system glasses, and they have argued that the differences in the values between the glasses and melts are insignificant [but Richet & Bottinga (1985) argue otherwise]. The partial molar entropies ($\Delta_m S_{di}^{liq}$) have proven to be the most difficult to estimate. This is because they cannot be directly measured experimentally. The approach has been to calculate actual partial molar entropies using all of the phase equilibrium experimental data, and then to find mixing models which fit the entropy values (Weill *et al.* 1980). As new experimental or thermochemical data have become available it has been possible to refine the entropy model (Henry *et al.* 1982). As can be seen in Figure 7.1, few compositions were determined by either Bowen or Murphy in the an-rich part of the ternary. This lack of liquidus data made it impossible to calculate and model partial molar entropies in that region. New experiments on an-rich compositions were recently done using Pt sample capsules sealed by welding to prevent sodium loss by volatilization (Zimmerman *et al.* 1985). These experiments have allowed a more refined entropy model to be constructed. This is a good example of how experimental design is guided by thermodynamic models.

7.2.2 Basaltic rock systems

Studies in simple systems such as di–ab–an can yield much information on the crystallization of specific minerals, but do not include all of the minerals important in the evolution of basaltic rocks. Experiments using an actual rock sample as the starting material can be used to obtain the crystallization behavior of all important minerals. Jumping from a simple ternary system to an eight or nine component basalt composition may cause experimental problems, all related to the presence of iron in the system. They are: iron loss to sample containers, control of f_{O_2}, and sodium volatilization. For atmospheric pressure experiments the use of Fe-saturated, Pt wire loops solves the Fe loss problem and allows rapid equilibration with the gas mixture (Donaldson 1979). However, Na loss can be at least 50% relative using the wire loop technique under reducing conditions (Grove & Bryan 1983). Sealed capules which would eliminate Na loss cannot be used if the f_{O_2} is externally controlled using gas mixtures. There is some evidence that Na loss is greater in H_2–CO_2 mixtures than it is in CO–CO_2 mixtures, presumably because Na volatility is enhanced by the presence of H_2O in the gas mixture (Biggar 1981). Even though Na loss has prevented the perfect experiment in atmospheric pressure basalt melting studies, much detailed information on the evolution of basaltic magmas by fractional crystallization has been obtained.

A good example of experiments to determine the liquid line of descent are those of Walker *et al* (1979) done on two selected mix-ocean ridge basalts (MORBs). Their experiments were begun with a careful choice of starting material based on detailed petrographic and geochemical studies of the samples. The experiments were designed to test two conflicting hypotheses for the chemical variation observed in basalts of the Oceanographer Fracture Zone of the Mid-Atlantic Ridge. Experiments were done using Fe-saturated Pt wire loops and at f_{O_2} values close to those of the quartz–fayalite–magnetite buffer, which is the f_{O_2} estimated for the MORBs. They analyzed (by electron microprobe) quenched glasses from runs containing plagioclase (Pl), olivine (Ol), and either high- or low-Ca pyroxene (Px), and plotted the glass analyses on a projection, as shown in Figure 7.2. The line shown in the figure indicates the path a multiply saturated (with Pl, Ol, and Px) liquid would take with falling T. For any given bulk composition, increasing T causes Px then Ol and then finally Pl to melt. Walker *et al.* extended the multiply saturated range to higher T by using diopside crucibles to contain the basalt sample and adding Pl and

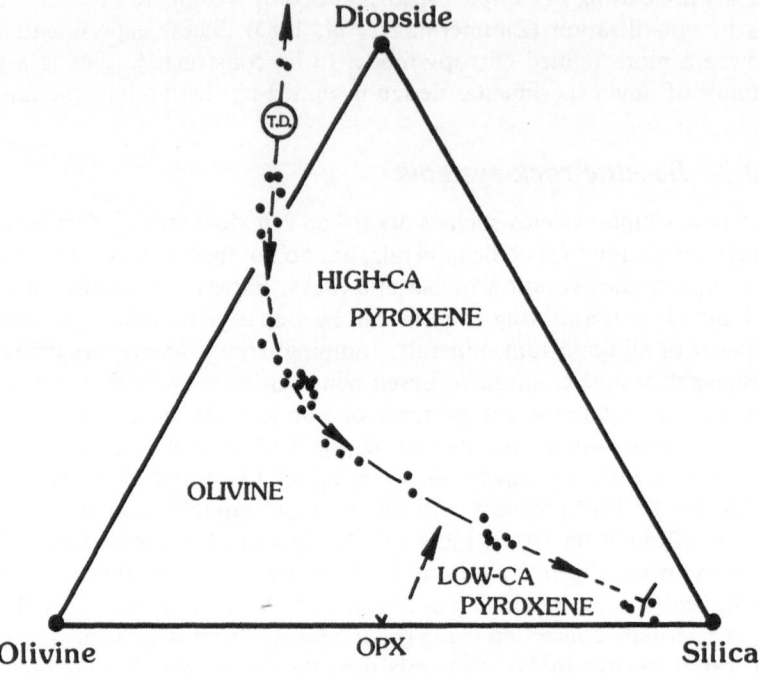

Figure 7.2 Plagioclase saturated liquidus projected onto the diopside–olivine–silica plane. Experiment at 1 bar with f_{O_2} controlled at QFM conditions. Plotted on a molecular proportion basis. Circled "T.D." is the thermal divide. The experimentally determined points shown were obtained by microprobe analysis of glasses quenched from runs saturated in three crystalline phases (high Ca–pyroxene + olivine + plagioclase or low Ca–pyroxene + high Ca–pyroxene + plagioclase). Taken from Walker *et al.* (1979).

Ol crystals to the basalt. Because Px and Pl crystals do not completely equilibrate with the melt in the run times of these experiments, these must be looked on as "quick and dirty" results. Nevertheless they provided the information the investigators sought.

Experiments in basaltic systems can be evaluated using several models for mineral/melt equilibria which are available. These are either empirical models aimed at predicting crystal/liquid equilibria in basaltic systems (Ford *et al.* 1983; Nielsen & Dungan 1983), or theoretically based. Ghiorso & Carmichael (1980) constructed a simple thermodynamic framework into which to fit the experimental data. This last approach is inherently the most powerful, providing there are sufficient data available and that the model is accurate enough to represent the data. It should be emphasized that the experimental data used in the models should be carefully selected, using criteria such as demonstrated reversibility and analytical accuracy.

7.3 Crustal magmas

7.3.1 Objectives

The experiments discussed in this section are aimed at unraveling the origin and evolution of the granitic magmas which dominate igneous rocks in the continental crust. We separate these magmas for discussion because one of their primary features is that they contain significant amounts of H_2O. The H_2O contents, together with their high SiO_2 activities result in lower melting range temperatures than found in basalts. The H_2O content, T and P range necessary for the study of these magmas fits ideally into the range of argon-medium, internally heated vessels (Section 3.4).

Some of the goals of experiments relevant to granitic rocks are

(a) to determine the ranges of P, T, and composition over which various minerals are stable in the melting range;

(b) to use these mineral stabilities to estimate the conditions of P, T, and H_2O content which occurred during the origin, ascent and emplacement of granitic magmas;

(c) to develop thermodynamically based models for crystal/liquid equilibria which can be used in conjunction with fluid dynamic models to describe the ascent path, rheology, texture, and emplacement behavior of granites.

7.3.2 The granite system

As late as the 1950s many geologists had the view that large granite batholiths formed by solidstate diffusion processes (e.g., Read 1948). In the first large-scale set of experiments done on the melting of a hydrous system Tuttle & Bowen (1958) conclusively demonstrated that granites formed by processing involving a silicate melt. Their experiments were developed over a period of years and demonstrate many excellent techniques. Their results are widely known and will not be described here, but some of the details of their approach are worth emphasizing. Their goal was to determine the crystal/liquid equilibria in H_2O-saturated mixtures of albite, sanidine, and quartz. They used an excess of H_2O, encased along with the silicate composition, in platinum capsules. The experiments were done over the pressure range 0.5 to 4 kbar.

Tuttle & Bowen were careful to explore the details of the system. For instance, they carefully characterized all of the crystalline phases, the melt (by measuring the solubility of H_2O in the quenched glass), and even attempted to describe the composition of the silicate component of the aqueous fluid phase. They noted that large ratios of H_2O : silicate resulted in differential leaching of the silicate, and so used only about 20 wt% H_2O in

111

the sample capsules. For the solidus determinations they used precrystallized starting material, but they noted that equilibration times varied with composition. The liquidus determinations were done using partly or completely crystalline material. However, they found that large-grained crystals took long times to equilibrate, and so kept the grain size of the starting material very small. They were able to do this because they observed that the final grain size of the crystals was proportional to the particle size of the initial glass.

It is worth noting Tuttle & Bowen's experiments were able to reach equilibrium even though temperatures were relatively low ($\sim 700°C$) and run durations relatively short (hours to weeks). The rapid attainment of equilibrium is due to the rapid Na−K interdiffusion in the alkali feldspars, and the alkalies are the only compositional variables in the crystalline phases. The situation with plagioclase is much different, as we will see in a later section.

7.3.3   Granodiorite systems

The majority of "granitic rocks" have more complex compositions than that of the simple granite systems studied by Tuttle and Bowen (1958). The major additions are CaO, which results in significant amounts of plagioclase, and FeO and MgO which result in biotite, hornblende, and pyroxene. In addition, it is now believed that most granitic magmas are undersaturated in H_2O. Indeed, as we noted above, one of the important reasons for experiments on granitic systems is to estimate the H_2O content of magmas. We address each of these complexities in later sections.

7.3.3.1   Plagioclase/liquid equilibria

The presence of anorthite in a system allows plagioclase to exist in addition to alkali feldspar. In his study of crystal/liquid equilibrium in the ab−an−H_2O system, Johannes (1978) showed that the addition of plagioclase to a system can cause severe experimental difficulties.

Johannes' starting materials consisted of synthetic plagioclase. The intermediate plagioclase was synthesized from glasses in the presence of H_2O for 2 days at 650°C and 2 kbar. The average grain size of these crystals was very small, < 1 μm. All of the equilibration experiments were done at 5 kbar, in the presence of an H_2O fluid. He found that, at 1,000°C, plagioclase compositions could be reproduced in run times of 1 h. Longer times caused no further change in plagioclase composition. However, in experiments just 100°C lower, at least 200 h were required to achieve constant composition plagioclase. Furthermore, in these 900°C experiments, the starting plagioclase composition had to be within 30% of the final composition or equilibrium could not be reached. In experiments at 800°C equilibrium was not even beginning to be approached in runs of 1,000 h. Johannes estimates that times as long as 100,000 h would be necessary to reach plagioclase/melt equilibrium at 800°C. Because many natural granitic rocks have solidus temperatures well below the 850 to 800°C range there is a great need to know crystal/liquid equilibrium relations in the experimentally inaccessible region. The only obvious way to solve the problem is to use an accurate thermodynamic model. One such model is described in the next section.

7.3.3.2   The Burnham quasi-crystalline melt model

Burnham (1975, 1979a, b, 1981) has developed a thermodynamic model to calculate quartz, alkali feldspar, and plagioclase melting relations in silicic magmas under crustal conditions. The original basis of the model was the pressure-volume-temperature measurements of the albite-H_2O system (Burnham & Davis 1971, 1974). The approach used in the model is to determine activities of silica and the feldspar components from experimental data on melting of the pure end-member minerals under dry and H_2O-saturated conditions, and then to determine interaction parameters for components in the melt based on melting relations in systems such as $SiO_2-NaAlSi_3O_8-H_2O$. The current version of the model (Burnham & Nekvasil 1986, Nekvasil & Burnham 1987) accurately reproduces the phase diagrams of Tuttle & Bowen (1958) as shown in Figure 7.3. The model allows calculation of crystal/liquid equilibria under the H_2O-undersaturated conditions found in most natural situations. The model does not require data on plagioclase/liquid equilibria measured at temperatures $<900°C$, but can be used to calculate crystallization and melting relations in the low T range (600–900°C), thus circumventing the experimental problem.

The Burnham model requires further testing on natural rock compositions such as andesite and granodiorite, especially for H_2O-undersaturated conditions. Its utility could be greatly increased by extending it to include pyroxene components, and then all other components important in granitic magmas. Experiments should be designed specifically to collect the necessary data. For instance, H_2O-undersaturated experiments on granodiorite compositions need to avoid the problems of disequilibrium in plagioclase. They also need to provide accurately measured compositions of coexisting plagioclase and melt.

7.3.3.3   Hydrous mineral melting

Biotite and hornblende are among the most important minerals in granitic rocks because they are reservoirs for H_2O and many incompatible trace

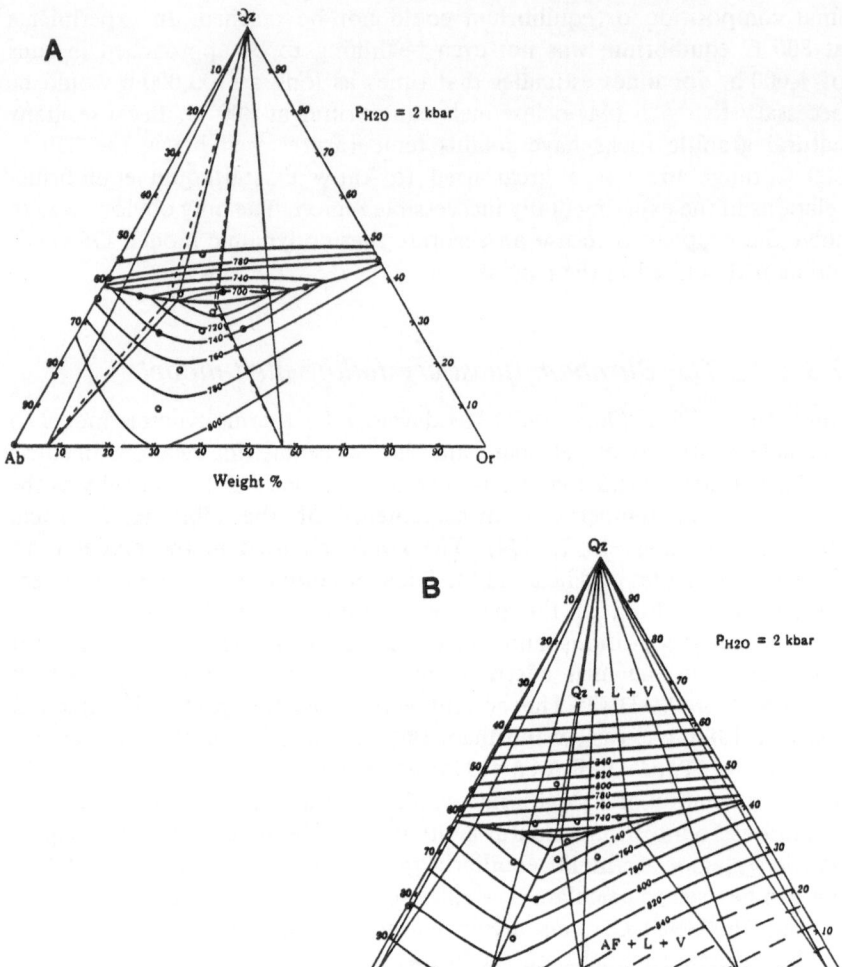

Figure 7.3 Comparison of experimentally determined (A) and calculated (B) liquidus phase relations in the system albite–orthoclase–quartz–H₂O From Nekvasil and Burnham (1987).

elements. Early experiments on these minerals in natural rock compositions gave confusing results, but suggested that their stability was complexly related to the activity of H_2O in the system. In cases where the results from natural rock systems are difficult to interpret it is often useful to investigate a simplified system. Holloway (1973) chose pargasite $[NaCa_2Mg_4Al_3Si_6O_{22}(OH)_2]$ as a simplified representative of natural hornblendes, and determined its melting relations as a function of pressure and H_2O activity.

The experiments were done in an argon gas medium, internally heated pressure vessel. The activity of H_2O was varied by using mixtures of H_2O and CO_2 generated from oxalic acid dihydrate or anhydrous oxalic acid. An example of the experimental results is shown in Figure 7.4. In order to demonstrate equilibrium, three types of starting material were used. Type I was a gel of pargasite composition, Type II a mixture of glass, olivine, clinopyroxene, and spinel crystals having the bulk composition of pargasite, and Type III a mixture of equal amounts of Type I and crystalline pargasite. These starting materials were sealed inside silver$_{70}$–palladium$_{30}$ capsules with a 1 : 1 ratio of silicate : fluid phase. The large amount of fluid prevented hydration–dehydration of pargasite or hydration of the melt from changing the $H_2O : CO_2$ ratio of the fluid during the run, so the ratio was known from the amounts of materials weighed into the capsule. The experiments were of three types. Synthesis runs were done using the Type I starting material to establish the approximate position of the melting reaction. However, because pargasite forms rapidly from the gel starting material as the experiment is being heated up, and pargasite apparently persists to temperatures above its stability limit, the results of these experiments overestimated the melting T. To overcome this, and demonstrate equilibrium, reversal experiments were done by first holding the run at a T known to be above

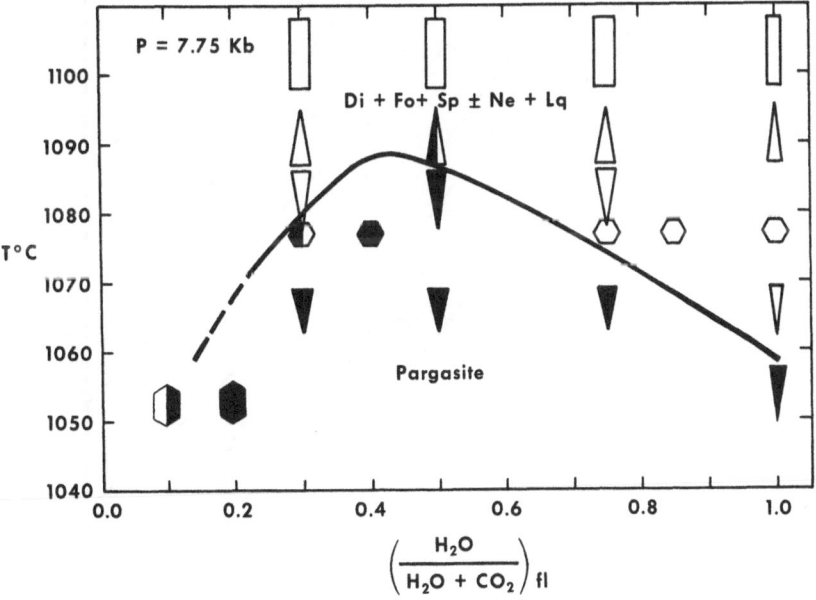

Figure 7.4 Melting of the hornblende mineral pargasite as a function of H_2O activity (proportional to $H_2O : H_2O + CO_2$ in the fluid. Symbols are: Type I, open rectangles; Type II, hexagons; and Type III, triangles (see text).

that of the melting reaction, and then reducing T to a chosen value. If pargasite was present at the end of that run, a true reversal had been accomplished. Unfortunately, the reversal runs did not react quickly when the T was close to the reaction boundary. To determine the direction of reaction over a small T interval the Type III starting material was used and the direction of reaction determined by measuring the relative intensities of X-ray diffraction peaks of pargasite and clinopyroxene. Because the temperature of the melting reaction was precisely determined as a function of the fluid composition, the results demonstrate the effect of H_2O activity (which is nearly equal to the mole fraction of H_2O in the fluid) on pargasite stability.

7.3.3.4 Minor mineral stabilities

Certain varieties of granitic rocks contain several minerals in minor abundances. Experimental determination of the stabilities of those minerals

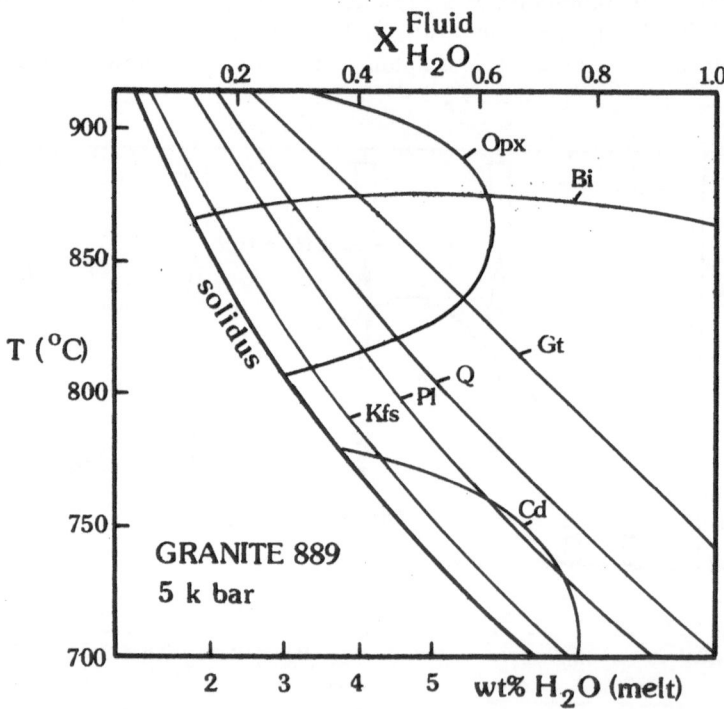

Figure 7.5 Phase diagram for an S-type granite at 5 kbar showing phase stabilities as a function of melt H_2O content or mole fraction of H_2O in the fluid. Phases are stable to the low H_2O content side of the lines except for biotite, which is stable at temperatures below its boundary. Taken from Clemens and Wall (1981).

often allow estimates of the $P-T$ conditions and H_2O contents of the melt to be made. The S-type (sedimentary source region) plutonic and volcanic rocks often contain orthopyroxene, cordierite, and garnet in addition to the major minerals.

Clemens & Wall (1981) determined the phase equilibria of an S-type volcanic rock as a function of P, T and H_2O activity. They used starting materials consisting of glass made from the powdered rock and then seeded with cordierite and garnet or, for near liquidus experiments, the rock powder itself. It was necessary to add the cordierite and garnet seeds because those minerals do not nucleate in the times used in the experiments. Figure 7.5 shows the results of their 5 kbar experiments. Note the complex nature of the phase boundaries, and their dependencies on the melt H_2O contents. Using this diagram and the observed phenocryst assemblage in the rock, Clemens & Wall (1984) were able to estimate that the magma producing this rock had a melt H_2O content between 2.4 and 4 wt% and began crystallizing at about 850°C at about 4 kbar.

7.4 Experiments at mantle pressures

Compared to the experiments discussed above, crystal/liquid experiments relevant to mantle magmas are more difficult due to the greater pressures required. The pressures are beyond the range of common gas vessels, and so experiments are restricted to the piston-cylinder apparatus up to 50 kbar and to more exotic pressure systems at higher P. These systems are much more limited in sample volume (Ch. 3), which makes control of f_{O_2} and volatile activities difficult. In addition the problem of iron loss to the sample capsule is greatly increased. For these reasons, very few experiments combining controlled volatile activities with iron-bearing systems have been done successfully.

7.4.1   The join diopside–forsterite–anorthite

Experiments on volatile- and iron-free systems are relatively straightforward, requiring only calibration of P and T in the high P system, and skill in execution and interpretation of results. As in the case of low P experiments, phase relations in simple systems at high P provide powerful constraints on the interpretation of more complex systems.

An excellent example of careful work on this simple system is the ongoing study of Presnall and coworkers (Presnall *et al.* 1978). Their experiments are designed to determine the effect of P on the compositions of partial melts in a spinel lherzolite mantle.

117

7.4.2 Experiments on volatile-containing systems

Eggler (1975a) demonstrated that CO_2 had significant solubility in basic melt compositions at pressures around 30 kbar, and that the presence of CO_2 in the melt changed the liquidus phase relations compared to volatile-free compositions. The change in phase relations indicated changes in melt composition and might explain the origin of highly basic magmas such as nephelinites, kimberlites, and carbonatites. Experiments applicable to such magmas thus require control of CO_2 activity.

In a comprehensive study of a number of joins in the $Na_2O-CaO-Al_2O_3-MgO-SiO_2-CO_2$ system, Eggler (1978) was able to demonstrate the major effect CO_2 can have on mantle melting relations. Starting materials were chosen to produce a pure CO_2 fluid. They were either mixtures of glass and silver oxalate ($Ag_2C_2O_4$), added to generate CO_2, or mixtures of oxides and carbonates, with the carbonates providing a source of CO_2. The samples were sealed inside Pt capsules, but any hydrogen present in the furnace assembly outside the capsule could diffuse into the capsule (Ch. 2) and react with CO_2 in the fluid to produce CO and CH_4. To cope with this problem, Eggler used furnace assemblies carefully dried at high T. He analyzed the quenched fluid by gas chromatography and showed that less than 1 mol% of the fluid was CO, and that the amount of CH_4 was negligible.

A few of Eggler's results are shown in Figure 7.6, with results from the volatile-free composition shown for comparison. It can be seen that CO_2 causes a dramatic freezing point depression between 25 and 30 kbar and that the carbonate (dolomite) becomes a stable phase on the solidus. In most of the experiments above 25 kbar, carbonate crystals formed on the quench together with quench pyroxenes. In many cases there was no trace of glass left to indicate the presence of a melt. The carbonates could be distinguished from their appearance, the quench crystals appearing cloudy and patchy, while stable carbonates were clear and granular (Irving & Wyllie 1975). The quench pyroxenes looked very much like stable ones, but could be distinguished by microprobe analysis. The quench pyroxenes had highly variable compositions falling inside the Opx–Cpx miscibility gap.

On the basis of the liquidus phase equilibria, Eggler postulated that CO_2 dissolves in relatively basic melts by the following reaction:

$$2SiO_4^{4-} + CO_2 \rightleftharpoons Si_2O_7^{6-} + CO_3^{2-}$$

$$\text{liq} \qquad \text{fluid} \qquad \text{liq} \qquad \text{liq}$$

This reaction involves polymerization of the melt, and explains the observation that addition of CO_2 to the melt favors crystallization of Opx instead of Ol.

118

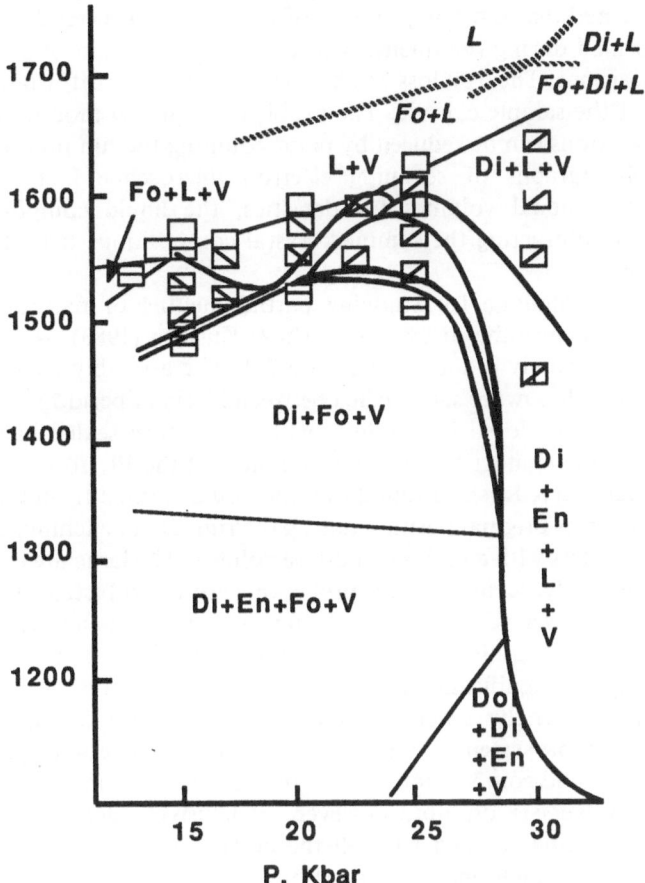

Figure 7.6 Vapor-saturated phase relations for the composition $(CaMgSi_2O_6)_{60}$ $(Mg_2SiO_4)_{33}(SiO_2)_7$ wt% in the presence of CO_2. Volatile–absent system shown as dotted lines. Di = diopside, Fo = forsterite, En = enstatite, Dol = dolomite, L = liquid, V = vapor. (From Eggler 1978).

7.4.3 Experiments on natural basalt–peridotite systems

Beginning with Yoder & Tilley (1962), and Green & Ringwood (1967), there have been many experiments on basaltic and peridotitic natural compositions in the 10–30 kbar range. As more of these experiments were done it was realized that many of the early results were seriously in error due in part to the old culprit, iron, but also due to large changes in liquid composition during the quench. Green (1976) presents a detailed analysis of the problems, showing that there were large losses of iron to the metal capsules as well as large increases in the ferric : ferrous ratios of the run products. He also documented the common presence of rims on stable crystals which he inferred formed during the quench and caused large modifications of the melt compositions. The iron loss can be greatly reduced by diffusing Fe into the inside of the sample capsule. The problem of quench modifications to glass compositions can be reduced by point-counting the run product using enlarged photographs of scanning electron microscope images. After correcting the modal volumes for densities, the liquid composition is calculated by subtracting the summed crystal compositions from the bulk composition.

A different technique for studying partial melting of natural mantle assemblages was introduced by Takahashi & Kushiro (1983). A sketch of their capsule assembly is shown in Figure 7.7. The assembly consists of a thin disk of basalt powder sandwiched between layers of peridotite powder. The sandwich is enclosed in graphite, which is in turn sealed in Pt. The graphite prevents contact between the sample and the Pt, so no iron loss occurs. Takahashi & Kushiro found that the liquid formed from the basalt powder stays in its original position during the run. On quenching, the melt quenches to a glass with a large crystal-free volume. The large areas of glass can be easily analyzed by electron probe and are not affected by quench rims on crystals because the crystals are far away from the analyzed glass. They demonstrated equilibrium between crystals and melt by using basalts which initially had significantly different compositions, and by showing that the resulting glass analyses were essentially identical. [Fujii & Scarfe (1985) modified the sample system to have two disks of basalt in between peridotite layers so that liquid compositions can be reversed in one run.] These liquid composition reversals do not, however, conclusively demonstrate that equilibrium is being reached with all the crystalline phases. That would require reversal of each phase composition.

Use of graphite is an effective way of preventing iron loss, but care must be taken that the starting material does not contain Fe^{3+}. If it does, the graphite will reduce the ferric iron to ferrous and produce CO_2, which can dissolve in the melt and alter the phase relations as seen in the previous section. In fact, the glass from experiments such as these should be analyzed for CO_2 using micro infrared spectroscopy.

The results of Takahashi & Kushiro (1983) and Fujii & Scarfe (1985) provide the best estimates to date of the compositions of basaltic magmas formed by partial melting of a volatile-free, spinel lherzolite mantle. Preliminary melting experiments of mantle phase assemblages at pressures up to 140 kbar have been done by Takahashi (1985). At the present time those very high pressure melting experiments are subject to very large T gradients across the length of the sample, but the experiments show that it will be possible to explore melting relationships in the deepest part of the upper mantle.

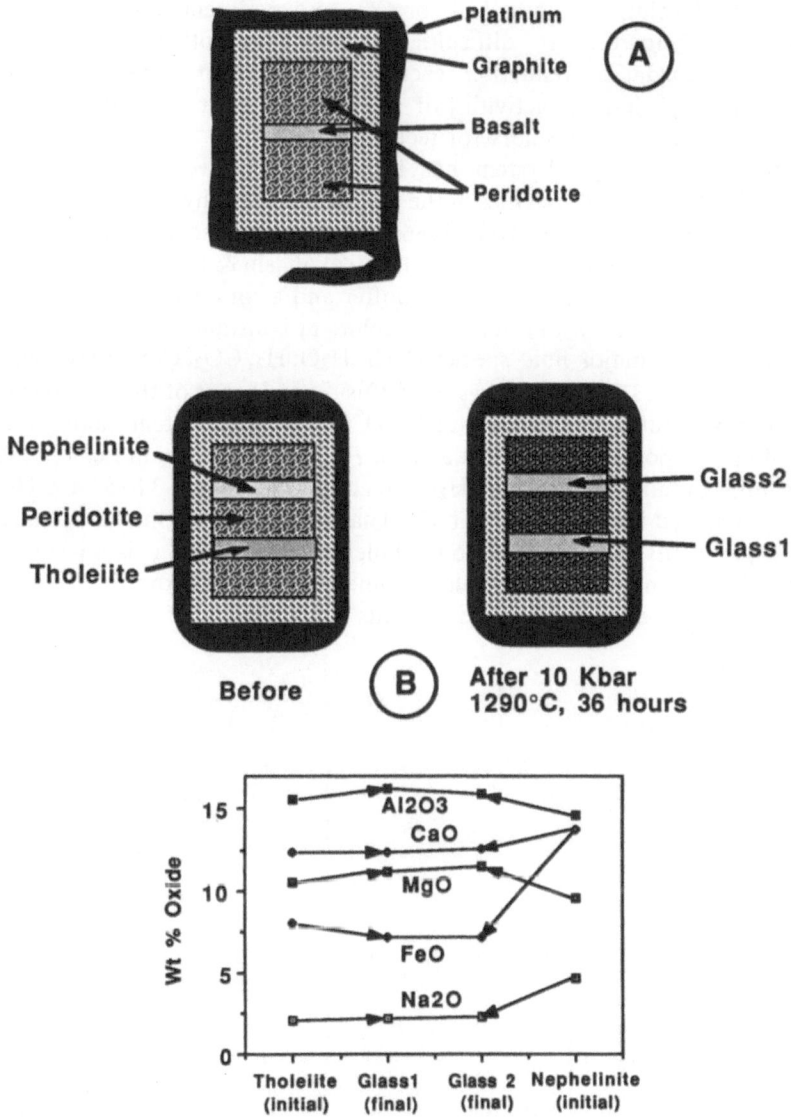

Figure 7.7 (A) Cross-section of the sample arrangement used by Takahashi and Kushiro (1983). The layer of basalt is about 1 mm thick. (B) Schematic cross-sections of before-run and after-run sample capsules used by Fujii and Scarfe (1985). The compositional data show that most elements in the glasses have converged on the same composition (glass1 and glass2) indicating significant reaction with peridotite minerals.

121

7.4.3.1 Experiments on volatile-containing mantle systems

Addition of volatiles such as H_2O or CO_2 to experiments on iron-bearing samples really increases the difficulties faced by the mantle experimentalist. Because of oxidation–reduction reactions between iron in the sample and the volatile species, the activities of the volatiles must be controlled (see Section 5.2 and the Appendix for techniques).

One sample capsule system has been found to work very well for controlling all volatile activities in the carbon–oxygen–hydrogen (C–O–H) system (Jakobsson & Holloway 1986). A sketch of the system (Fig. 7.8) shows it to be much like the one used by Takahashi & Kushiro (1983) with the addition of the iron–wustite f_{O_2} buffer and a source of C–O–H fluid. The combination of f_{O_2} buffer and graphite at constant P and T fixes the activities of all major fluid species (CH_4, H_2O, H_2, CO_2, CO). Pyrex glass surrounding the Pt capsule slows the diffusion of H_2 out of the capsule and allows run durations of 12–24 h at $1,200°C$. With this system there is only one fluid composition possible at a given P and T, as shown in Figure 7.8b. The fluid contains 10–20 mol% H_2O which results in about 3 to 5 wt% H_2O being dissolved in the coexisting melt. That amount of H_2O is far too large for experiments to be applicable to tholeiitic basalts, but it is relevant to basic alkaline magmas and to calc-alkaline magmas. The low f_{O_2} keeps the ferric : ferrous ratio very low and prevents the formation of spinels contain-

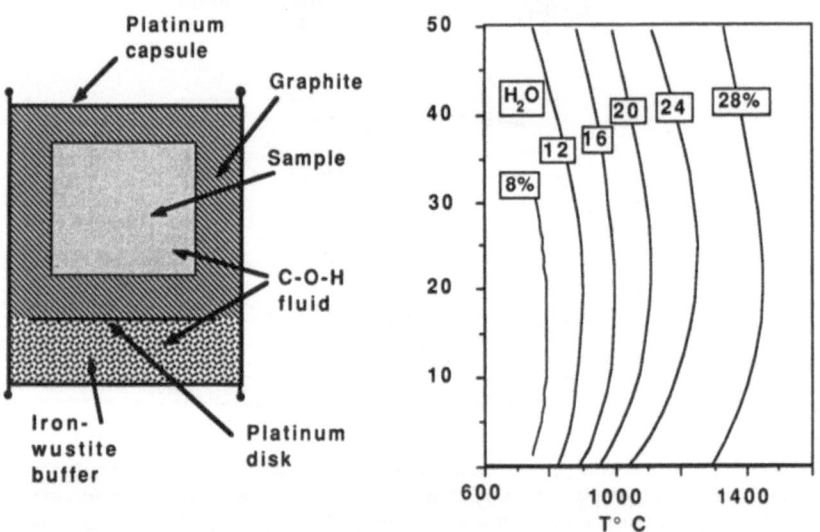

Figure 7.8 Cross-section of the capsule assembly for controlled activities of H_2O, CH_4, H_2, CO, and CO_2 in the fluid, after Jakobsson and Holloway (1985). P–T projection of the mole fraction of H_2O in the fluid buffered by iron + wustite + graphite (calculated as described by Holloway 1987).

122

ing ferric iron. The utility of the technique is that it allows well-controlled experiments which can provide accurate data on complex systems. The data from these experiments will be valuable for testing thermodynamic models of mantle melting.

7.5 Trace element behavior

Techniques for the determination of the trace or minor element contents of rocks and minerals first became available nearly 40 years ago. Since that time geochemists have attempted to use increasingly precise trace element data to understand the processes involved in the origin and evolution of igneous rocks. The reason for the importance of trace elements in this regard is that most of the 91 naturally occurring elements are present in tiny quantities while only eight or nine are present in amounts greater than about 1 wt%. The large number of trace elements exhibit great ranges in geochemical behavior, from strong tendencies to enter crystallizing minerals (compatible) to essentially 100% retention in the silicate melt (highly incompatible). In addition, some have pronounced affinities for certain crystal structures, but not for others. Thus, if one were to determine the trace element contents of a suite of genetically related lavas, it should be possible to establish the different extents of initial partial melting and later crystallization which resulted in the final products.

The evolution of a magma body is generally modeled by considering the effects of the following processes:

(a) equilibrium partial melting;
(b) fractional crystallization involving one or more minerals; and
(c) mixing or contamination.

These processes, and one or two variants, may be represented by simple analytical expressions or modeled numerically by computer. In order to quantify their different effects, however, it is necessary to know how each of the trace elements is partitioned between the melts and the different minerals which crystallize from them. The partition coefficient $D^{A/B}$, describing the distribution of element i between phase A and phase B, is defined as follows:

$$D^{A/B} = \frac{\text{concentration of } i \text{ in phase } A \text{ (by weight)}}{\text{concentration of } i \text{ in phase } B \text{ (by weight)}}$$

Initially, values of $D^{\text{crystal/melt}}$ for different elements were derived by separating phenocrysts from their volcanic matrices, determining the trace element contents of both, and assuming quenched-in equilibrium. These values were then applied to petrogenetic arguments by assuming that D is

123

independent of the concentration of the trace element (Henry's Law) and that it does not vary substantially as the major element content of the magma changes.

Henry's Law is based on the principle that if a few atoms of a trace component are put into a host of fixed major element composition, then each trace atom is in an environment completely dominated by the host. Thus, the activity coefficient of the trace component should be independent of whether its concentration is 1 ppm or 100 ppm, because at these low levels the trace concentration does not affect the overall properties of the host. While this assumption is reasonable, it is clear that changing the major element composition of the phase must affect the environment of the trace substituent and hence will change its activity coefficient and partition coefficient with other phases. Partition coefficients should also depend on pressure and temperature in the same way that other equilibrium relationships do.

The main aims of experimentalists investigating crystal/liquid partitioning have been to test the concentration limits of Henry's Law behavior and to determine the effects of major element composition, temperature and pressure on trace element partitioning. Before considering some of the experiments in detail, however, it should be made clear that trace elements fall into at least two very distinct groups. There are those which, when entering crystals, substitute for an ion of the same charge (homovalent substitution), and those which enter lattice positions normally occupied by ions of different charge (heterovalent substitution). One might also consider a third group: those which enter interstitial positions or other positions not generally occupied by the major components of the mineral. Substitutions of this third type lead to values of $D^{crystal/melt}$ close to zero, however, and are difficult to characterize quantitatively, so they will not be considered further here.

7.5.1 Homovalent substitution

Trace elements such as Ni, Co, Mn, Sr, and Ba provide examples of possible homovalent substituents in silicate materials. In the cases of Ni, Co, and Mn, atoms in the $+2$ oxidation state generally occupy lattice positions normally filled by Mg and Fe^{2+} while Sr and Ba replace calcium. Considerable attention has been given to the behavior of Ni, which enters silicate minerals, particularly olivine, very readily and which is hence a sensitive indicator of crystal–liquid processes.

Drake & Holloway (1981) made a detailed study of the partitioning of Ni between olivine and silicate melt under conditions of fixed pressure and temperature. They used two bulk compositions $(CaMgSi_2O_6)_{70}(Mg_2SiO_4)_{25}(SiO_2)_5$ and $(NaAlSi_2O_6)_{80}(Mg_2SiO_4)_{20}$ and performed experiments at 1 atm pressure, $1,400°C$ and 1 atm, $1,300°C$, respectively. In order to perform experiments at low concentrations of Ni, the bulk compositions were spiked with radioactive ^{63}Ni in NiCl solution. The use of the β-emitting ^{63}Ni enables the concentration of Ni to be determined down to sub-ppm levels by the technique of autoradiography (Mysen & Seitz 1975). After the experiments are completed, polished sections of the charges are made and these sections placed in contact with a β-sensitive emulsion. The densities of β-tracks in parts of the emulsion in contact with olivine and glass are proportional to Ni concentrations in the two phases. The track densities are generally determined by counting or by microprobe analysis of the emulsion for silver.

For high Ni concentrations, up to 6 wt%, a small amount of the Ni was added to the charge as ^{63}Ni and the remainder as stable Ni in solid $NiCO_3$. The starting materials were turned into glasses by multiple fusions at 1,430°C for 1 h followed by grinding to fine powders. Multiple grinding and fusion is necessary to ensure isotopic homogeneity of the sample. The glasses were heated above the liquidus for 1–24 hours, then brought to run temperature and held there for about 4 days. In order to minimize the possibility of Ni loss to Pt capsule material, the charges were suspended on thin Pt wire loops; little or no Ni loss resulted. Partition coefficients were reversed by making runs in which the forsterite component was added as crystalline olivine containing either no Ni or all of the Ni.

Drake & Holloway found that it was possible to reverse the equilibrium value of D in 4 days with both bulk compositions. They also obtained the same partition coefficients from autoradiography and direct microprobe analysis under the high Ni concentration conditions where both could be used. Henry's Law, with constant D, was obeyed over the entire concentration range studied, 10 ppm to 60,000 ppm of Ni in olivine. Finally, $D^{ol/liq}$ was found to be highly dependent on major element composition, changing from 5 for $Di_{70}Fo_{25}Q_5$ to 18 for $Jd_{80}Fo_{20}$.

The observed dependence of $D^{ol/liq}$ on bulk major element composition and adherence to Henry's Law is consistent with the observations of Watson (1977) and Hart & Davis (1978). The former effect may be explained (Watson 1977) by considering the exchange of Ni with the ion it replaces, Mg, in olivine and melt:

$$NiO + MgSi_{0.5}O_2 \rightleftharpoons MgO + NiSi_{0.5}O_2$$

melt   olivine   melt   olivine

The equilibrium constant for this reaction is defined as follows:

$$K = \frac{a_{MgO}^{melt} \cdot a_{NiSi_{0.5}O_2}^{ol}}{a_{MgSi_{0.5}O_2}^{ol} \cdot a_{NiO}^{melt}}$$

Taking account of the fact that the olivine and melt obey Henry's Law (constant activity coefficient) at these concentrations and that the olivine is almost pure forsterite ($a_{MgSi_{0.5}O_2}^{ol} = 1$) this becomes:

$$K' = a_{MgO}^{melt} \cdot \frac{[Ni^{ol}]}{[Ni^{melt}]}$$

where the bracketed terms are concentrations instead of activities. If the activity of MgO in the melt is proportional to its mole fraction, then this expression may be rearranged to give:

$$D_{Ni}^{ol/melt} = \frac{constant}{X_{MgO}^{melt}}$$

Therefore, one would expect $D_{Ni}^{ol/melt}$ to be inversely proportional to the MgO content of the melt. Hart & Davis (1978) found this type of relationship to hold very well for Ni distribution between forsterite and a range of melts. They suggested that most of the variation in D from one experimental system to another was due to this compositional effect and that the influence of temperature was small with respect to it. In terms of weight percent MgO, Hart & Davis give the following equation for $D^{ol/melt}$

$$D_{Ni}^{ol/melt} = \frac{124}{MgO} - 0.9$$

Drake & Holloway (1981) found that their reversed data were in reasonably good agreement with predictions based on this simple equation.

In conclusion, it appears that homovalent substitution of Ni for Mg results in partition coefficients which are independent of Ni concentration (Henry's Law) over the range of concentrations observed in nature. Most of the variation in D from one system to another can be explained in terms of major element variations with temperature of crystallization being relatively unimportant.

7.5.2 Heterovalent substitution

Heterovalent substitutions, particularly of rare earth elements (REE), for Ca^{2+} and Mg^{2+} are extremely important in trace element geochemistry, because the lanthanides exhibit only slight differences in chemical behavior from one to another. Since these effects are systematic with increasing atomic number from La to Lu the ratios of light REE to heavy REE in igneous rocks are considered to be valuable indicators of petrogenetic processes. Most of the lanthanides have the 3 + oxidation state in nature so that their substitution for Ca^{2+} and Mg^{2+} is heterovalent. Crystallochemically, there are two ways that a 3 + ion can substitute for a

2+ ion and still preserve charge balance in the crystal. One is to couple the substitution with another heterovalent substitution. Lanthanides can, for example, make a charge balanced replacement of calcium if it is coupled with substitution of Al^{3+} for Si^{4+}, i.e., $CaSi \rightleftharpoons LaAl$. Another way is to substitute two lanthanides for three calciums and make a cation vacancy in what is normally a Ca position, i.e., $3Ca \rightleftharpoons 2La + V_{Ca}$.

Harrison & Wood (1980) have made a detailed study of the operation of these mechanisms by determining the partitioning of REE between garnet and hydrous silicate melt at 30 kbar pressure. Before describing their results in detail, it is useful to consider the probable concentration dependence of $D^{gt/melt}$ for the operation of the two mechanisms discussed above.

The coupled substitution mechanism may be represented by an equilibrium involving M^{2+} (the divalent host), REE^{3+}, Al^{3+}, and Si^{4+} as follows:

$$REE^{3+}_{melt} + Al^{3+}_{melt} + (MSi)_{gt} \rightleftharpoons (REE\ Al)_{gt} + M^{2+}_{melt} + Si^{4+}_{melt}$$

For experiments performed under fixed conditions of P, T, and major element composition, the concentrations and activities of M, Si, and Al in the melt are all fixed, as is that of (MSi) crystal so we have

$$K = \frac{a^{gt}_{(REE\ Al)} \cdot a^{melt}_{M^{2+}} \cdot a^{melt}_{Si^{4+}}}{a^{melt}_{REE^{3+}} \cdot a^{melt}_{Al^{3+}} \cdot a^{gt}_{MSi}}$$

$$K' = \frac{a^{gt}_{(REE\ Al)}}{a^{melt}_{REE^{3+}}} = constant \approx D^{gt/melt}_{REE}$$

Under conditions of low concentration of REE it seems likely that Henry's Law will be obeyed in both garnet and melt phases and that $D^{gt/melt}$ will be independent of REE concentration.

The defect substitution mechanism may be described by a similar equilibrium involving M^{2+}, REE^{3+} and V_m, the vacancies in M positions:

$$2REE^{3+} + 3M^{2+} = 3M^{2+} + 2REE^{3+} + V_m$$

$$\text{melt} \qquad \text{gt} \qquad \text{melt} \qquad \text{gt}$$

Taking account of the fixed activities of M^{2+}_{gt} and M^{2+}_{melt} the equilibrium constant is rearranged as before:

$$K' = \frac{(a^{gt}_{REE})^2 \cdot (a^{gt}_{V_m})}{(a^{melt}_{REE})^2}$$

Assuming, at high dilution, Henry's Law behavior for REE in garnet and melt and for vacancies V_m in garnet, activities may be replaced by

127

concentrations [ ] as follows:

$$K'' = \frac{[REE_{gt}]^2[V_m]}{[REE_{melt}]^2}$$

Therefore, the distribution coefficient $D^{gt/melt}$ in terms of concentration depends on the concentration of vacancies:

$$D^{gt/melt}_{REE} = \frac{(K'')^{1/2}}{[V_m]^{1/2}}$$

The vacancies present in the garnet are of two types. There are those formed as part of the heterovalent substitution (V_ε) and those which are intrinsically present in the pure crystal irrespective of the presence of REE impurities (V_i).

$$V_m = V_\varepsilon + V_i$$

For the defect substitution reaction we have:

$$V_\varepsilon = \tfrac{1}{2}[REE_{gt}]$$

If V_ε dominate over V_i the concentration of vacancies may be replaced by REE concentration:

$$D^{gt/melt}_{REE} = \frac{constant}{[REE_{gt}]^{1/2}}$$

and partition coefficient D clearly depends on rare earth element concentration so that Henry's Law would appear not to be obeyed. On the other hand, if V_i dominates, the vacancies are an intrinsic property of the crystal rather than substitution-controlled (Section 9.3). Under these circumstances, $D^{gt/melt}$ would be independent of the concentration of trace element in the crystal.

In summary, coupled substitution involving combined lanthanide and aluminum replacement of M^{2+} and silicon should result in $D^{gt/melt}$ which is independent of REE concentration in the trace element range. Substitution of lanthanide for M^{2+} coupled to vacancy production leads to constant $D^{gt/melt}$ if the intrinsic vacancy concentration dominates over that produced by the substitution. If the extrinsically produced substituent vacancies are more important, $D^{gt/melt}$ must depend on lanthanide concentration even in the trace element range. Intuitively, it seems probable that V_i dominates at very low concentrations of trace element, and that the extrinsically produced vacancies become more important as REE concentration increases.

Finally, the vacancy mechanism obviously can only work at low concentrations of V_m because the crystal cannot accommodate large numbers of highly charged holes in its structure. Thus, at high REE content, the coupled [REE Al] substitution must be the only important one. This conclusion is supported by the fact that completely pure REE Al garnets such as $Y_3Al_5O_{12}$ are stable at low pressures.

Harrison & Wood (1980) determined the partitioning of two lanthanides, Sm and Tm between pyrope ($Mg_3Al_2Si_3O_{12}$) and $Mg_3Al_2Si_3O_{12}-H_2O$ melts and grossular ($Ca_3Al_2Si_3O_{12}$) and $Ca_3Al_2Si_3O_{12}-H_2O$ melts at 30 kbar. Their results will be illustrated by discussing the observed partitioning of Sm in the $Mg_3Al_2Si_3O_{12}-H_2O$ system at 30 kbar and $1,300°C$. Glass of $Mg_3Al_2Si_3O_{12}$ composition was prepared by melting the oxides at $1,800°C$ and then quenching. After grinding the glass, β-emitting ^{151}Sm and stable Sm were added in dilute chloride solutions. The spiked glasses were ground again and melted at $1,500°C$ for 12 hours to homogenize the glass and expel all chlorine. Crystalline pyrope was synthesized from Sm-free glass at 30 kbar and $900°C$ using sealed capsules and H_2O fluid. Initial runs were made on the Sm-bearing glasses. These were sealed with water in platinum capsules and run for varying lengths of time at 30 kbar and $1,300°C$. The distribution of Sm between garnet and quenched melt was determined by autoradiography. Runs of increasing time up to about 5 hours produced progressively decreasing values of $D^{gt/melt}$, but no further decrease occurred in run times of from 5 to 20 hours. Harrison & Wood concluded that garnet initially crystallizes with too great a concentration of samarium and that the excess gradually diffuses out in runs of increasing duration. In order to obtain the equilibrium values of $D^{gt/melt}$ they bracketed their results by starting with a mixture of glass containing all of the Sm and garnet crystals with no Sm. These experiments yielded the same values of D as the synthesis runs in about 5 hours at $1,300°C$. Figure 7.9 shows the observed partition coefficient as a function of samarium concentration under these conditions. It can be seen that D is independent of the REE content of garnet in the range of 1 to 300 ppm but increases drastically as Sm is reduced from 1 to about 0.1 ppm.

The region of deviation from Henry's Law was interpreted by Harrison & Wood as the result of Sm plus defect substitution with substituent coupled vacancies dominating over intrinsic vacancies. The high concentration, Henry's Law, region was considered due to coupled SmAl = MgSi replacement. In order to test that the low concentration non-Henry's Law region is due to REE-generated vacancies, Harrison & Wood (1980) repeated the low concentration runs but added substantial amounts of several other REEs, lanthanum, dysprosium, and lutetium. If the hypothesis were correct, these should have affected $D^{gt/liq}$ by also generating vacancies and hence shifting the samarium partition coefficient down into the Henry's Law region. This is exactly what was found. In nature the defect substitution mechanism,

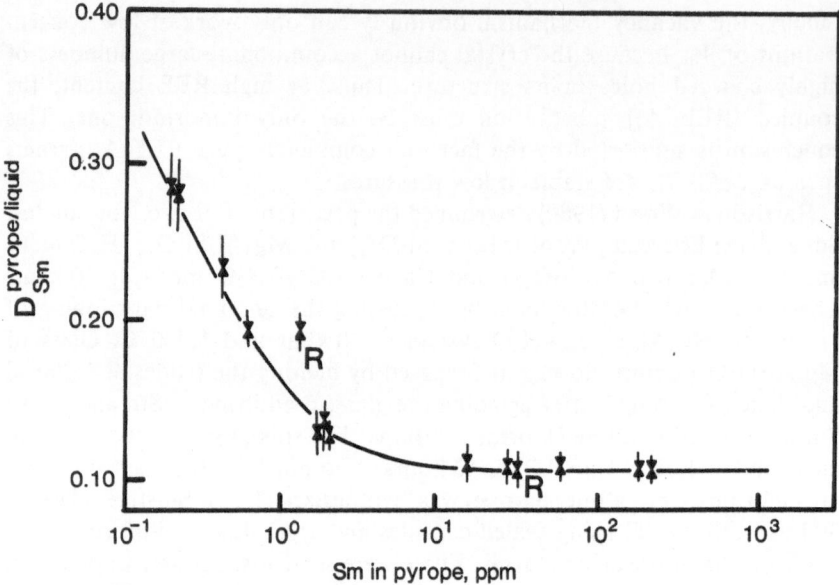

Figure 7.9 Partition coefficients for samarium between pyrope garnet and liquid at 30 kbar and 1,300°C as a function of samarium content of the garnets (ppm). R denotes a reversed run. Error bars ±1σ.

which operates up to a few ppm, should be saturated by the sum of all the REE present and Henry's Law, constant D_{REE}, behavior observed for each of them.

In conclusion, it appears that lanthanide substitution into garnet, and probably into other silicates too, exhibits departure from Henry's Law behavior at concentrations up to a few ppm. The mechanism is not an important influence on partitioning behavior at high REE concentration because the defect mechanism becomes saturated. Partitioning in natural systems can generally be approximated by coupled $(REE\,Al) = (M^{2+}Si)$ substitution which obeys Henry's Law and constant D over a wide range of lanthanide concentrations.

7.5.3 Bulk compositional effects

Although it has been demonstrated that bulk, major element contents of natural phases affect trace element behavior, the magnitudes of such effects are difficult to separate from those of other variables such as pressure and temperature.

Watson (1979b) made a study of the solubility of an important trace element, zirconium, in felsic liquid close to the minimum melting composition in the granite system. Although behaving as a trace element during most petrogenetic processes Zr precipitates as the major constituent of Zircon, $ZrSiO_4$, at very low concentrations. This gave Watson the opportunity to investigate the mechanism of dissolution of Zr in silicate melts. Experiments were performed on compositions with variable Na:K and alkali:Al_2O_3 ratios in order to determine the concentrations of Zr in the liquid under the conditions at which zircon precipitates. Runs were performed for 6 to 30 days at 700°C to 800°C and 2 kbar in conventional cold-seal pressure vessels. Under these conditions, with excess H_2O fluid, the melts contain about 6 wt% H_2O. Figure 7.10 shows a plot of maximum Zr content of the melt as a function of the total alkali/Al_2O_3 ratio. It can be seen that, if there is no excess of alkali over the 1:1 ratio with Al_2O_3, the maximum solubility of Zr in granitic melts is extremely low, less than 100 ppm. In peralkaline melts (total alkali/$Al_2O_3 > 1$), Zr solubility increases dramatically with increasing alkali:alumina ratio (Figure 7.10). On a molar basis, it was found that a 2 mol excess of Na_2O or K_2O is needed to dissolve one additional mole of ZrO_2. Zircon solubility was found to be unaffected by SiO_2 content in the range 75–84 mol% or by Na:K ratios between 0.6 and 1.6. The implication is that zirconium dissolves in felsic melts as a complex involving alkali and zirconium in the ratio $2(Na_2O + K_2O):1(ZrO_2)$. This complex is less stable than that involving

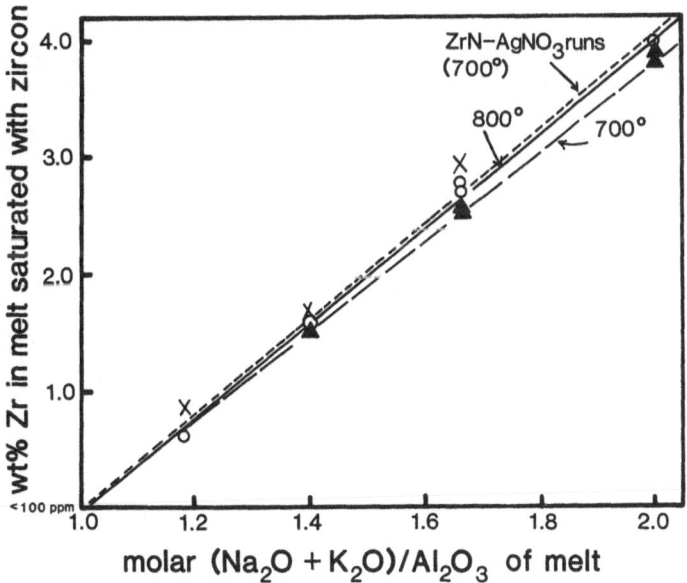

Figure 7.10 Weight percent dissolved Zr in melts saturated with zircon *vs.* molar (Na_2O + K_2O)/Al_2O_3 of melts. Circles and triangles represent runs at 800 and 700°C, respectively, in which Zr was added initially at ZrO_2. X shows values for 700°C runs in which Zr was introduced as ZrN (from Watson 1979b).

aluminum which has a ratio of $1(Na_2O + K_2O) : 1(Al_2O_3)$. Therefore, Zr only becomes substantially soluble in silicate melts when alkali : $Al_2O_3 > 1.0$. Watson used these data to model the behavior of Zr in metaluminous and peralkaline melts during partial melting and fractional crystallization processes.

7.6 Conclusions

We have seen that many good experiments have been done on igneous systems. These experiments have provided much insight into the origin and evolution of igneous rocks and such phenomena as volcanic eruptions. However, much is left to be done before we are able to answer important questions such as: At what depth do particular basaltic magmas originate, and how are their fractionation paths affected by P_{CO_2}? Very few data are available on the melting relations of mantle materials at pressures above 50 kbar, and thermodynamic models for silicate melts will have to be extended greatly before the effects of volatile components can be predicted over wide ranges of P, T, and composition. Advances in these areas will require new experimental apparatus and technique, and an even closer wedding of experiment and theory.

8 Igneous experiments on melts and fluids

8.1 Introduction

Volatiles play a central role in many magmas, changing melt structure, liquidus phase relationships, and the physical properties of melts. Separation of volatiles from magmas by formation of a fluid phase is the critical process in many volcanic eruptions and may play that role in formation of some ore-forming fluids. We have separated the discussion of experiments involving melt–fluid interaction into a separate chapter because many of the techniques are quite different from those used in crystal–liquid experiments. The objectives in the experiments described here are to provide data for thermodynamic models of volatiles in melts and to determine element partitioning between melts and fluids.

8.2 Volatile solubilities in melts

A volatile's effect on a melt depends on how much of it can dissolve in the melt, measured on a molar basis. The profound effect that a few wt% H_2O has on a melt is due to the low molecular weight of H_2O compared to the silicate units in melts. Measurement of volatile solubilities provides primary data for thermodynamic models of volatiles in melts. Solubility can also be used directly in modeling the mechanisms of volcanic eruptions.

Volatile solubilities are strong functions of P because the fugacities of volatile species in the fluid phase are strong functions of P (see Figure 8.1). Most solubility experiments consist of the equilibration of a melt with a coexisting pure fluid of the volatile in question, followed by analysis of the quenched glass for the volatile.

The solubility of H_2O in basaltic and andesitic melts was measured by Hamilton *et al.* (1964) In their study they evaluated four different techniques to determine solubility:

(a) The presence or absence of excess fluid pits in charges of known H_2O content.
(b) A modified Penfield tube method of direct H_2O analysis.
(c) Weight loss at $110°C$.
(d) Weight loss at $1,000°C$.

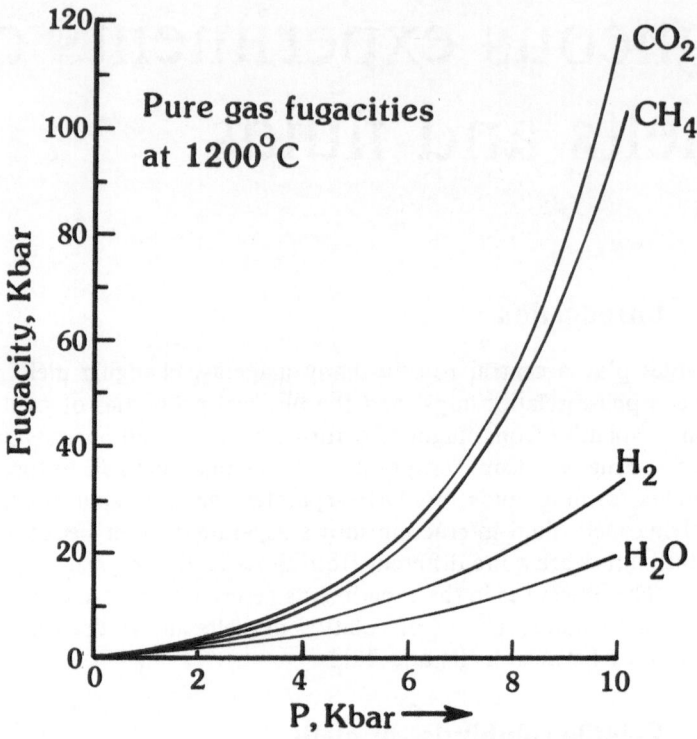

Figure 8.1 Fugacities of pure fluids as a function of P. Note the factor of 10 difference between the P and fugacity scales. Also note the large increase in fugacity of CO_2 and CH_4 above 6 kbar.

Method (a) had been used in earlier experiments on melts of low iron content (albite and granite, Burnham & Jahns 1962) but did not work well with the high-iron basalt. Hamilton *et al.* felt that this was because the oxidation of ferrous iron by H_2O during the run decreased the amount of H_2O available to dissolve, and that H_2 produced in the reaction also formed pits in the charges. Method (b) involved grinding the quenched glass to -80 mesh before analysis. The results using this method agreed with those from methods (c) and (d) at lower P but at higher P the H_2O contents determined with method (b) were much lower. This was attributed to loss of water in the grinding process. Method (c) measures the amount of H_2O remaining in the fluid, and so gives the amount dissolved as the difference between that weighed into the capsule and weight loss at $110°C$. The results from this technique were considered the best. The amount of H_2O used in oxidation of ferrous iron was calculated by analyzing the starting material and run product for ferrous and ferric iron. Method (d) was used as a check on method (c) and did yield consistent results. Because dissolved H_2O can react

134

with ferrous iron in the sample during the $1,000°C$ heating process, it was necessary to analyze the sample for ferrous iron before and after heating.

The experiments by Hamilton et al. (1964) point out most of the problems in solubility measurements. One of these problems is that of fluid inclusions in the sample. These can be formed during the run, or in high H_2O content samples (above about 5 wt%) during the quench. Inclusions are formed during the run if the starting material is a powder. In low-viscosity melts, these inclusions usually migrate out of the interior of the melt and are not a problem. In high-viscosity melts the inclusions remain in the melt. The number of inclusions, and consequently the apparent solubility, increases with decreasing grain size of the starting material (Tuttle & Bowen 1958, p. 14). Consequently, the best way to measure solubility in high-viscosity melts is to make a bubble-free glass of the starting material by vacuum melting, then break the bubble-free glass into millimeter size pieces, and saturate them by diffusing the fluid into them. This technique prevents formation of inclusions during the run so the only inclusions present will be those formed during the quench. Then in doing method (c) type analyses, it is important not to include weight loss from decrepitation of inclusions formed on the quench in high H_2O content (high P) runs. A good way to do this is to heat the sample in a thermal gravimetric balance in which weight loss is followed continuously as a function of time. For samples that do not form fluid inclusions on quenching, in situ analysis of H_2O or CO_2 can be done using micro-infrared spectroscopy or, in the case of H_2O, ion microprobe analysis for H. McMillan & Holloway (1987) present a more detailed discussion of H_2O solubility experiments.

Another little used technique that has great potential is to equilibrate the fluid with a cylinder of melt and measure the diffusion profile of the fluid species in the quenched glass. This method provided a few solubility measurements in a study of H_2O diffusion by Karsten et al. (1982). One of the diffusion profiles from their experiments is shown in Figure 8.2. The concentration data were obtained by ion microprobe analysis for hydrogen. A calculated profile was fitted to the data and extrapolated to the diffusion interface. The interface value gives the equilibrium solubility. Experiments of this type yield both diffusion coefficients and solubility in a single experiment.

Low viscosity melts at high P often cannot be quenched to glasses. In these cases, volatile solubilities must be determined indirectly. Order of magnitude estimates can be made from observation of freezing-point depression (Eggler et al. 1979). Phase equilibria can also be used to measure volatile solubilities quantitatively as done by Eggler (1975b).

The principles of the phase equilibrium technique are illustrated in Figure 8.3. The solubility of CO_2 and H_2O in enstatite and fluid-saturated melt is given by point N. Point N is determined by the intersection of the

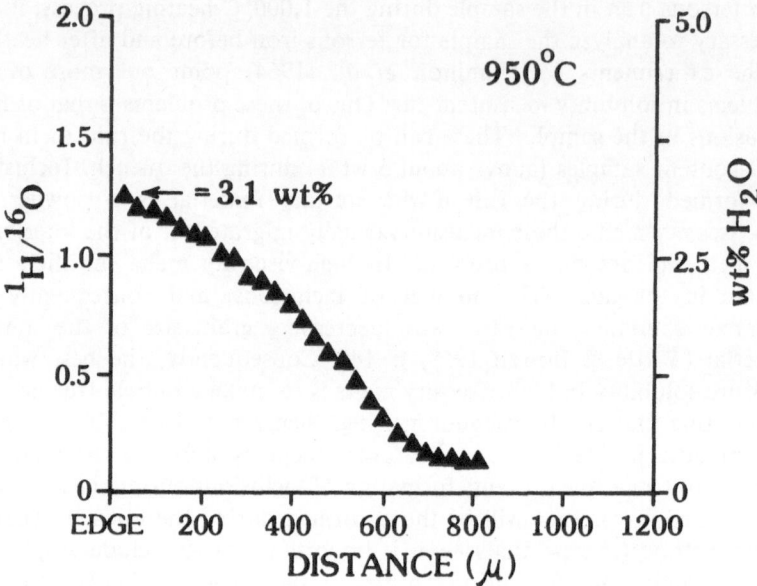

Figure 8.2 Diffusion profile of H_2O into rhyolite obsidian. The experiment lasted 3 hours at 700 bar and 950°C. The profile was determined by measuring H with an ion microprobe. Note the edge value which gives the equivalent H_2O solubility (shown on the right-hand ordinate scale). (Redrawn from Karsten *et al.* 1982.)

lines labeled B and C. Line C is determined by the enstatite apex and the experiments labeled d and e, which bracket the appearance of enstatite. Line B is determined by point M and the experiments f and g, which is also determined by the appearance of enstatite. Point M is fixed by line A which is given by the enstatite apex and experiments h and i, which is based on the appearance of liquid.

It can be seen that line A defines the $CO_2 : H_2O$ ratio in the fluid, line B defines the $H_2O : MgSiO_3$ ratio in the melt, and line C defines the $CO_2 : H_2O$ ratio in the melt. The precision of the technique is determined by weighing errors for loading $MgSiO_3$, CO_2 and H_2O into the capsule, and by the sensitivity of detecting small amounts of glass (line A) or enstatite crystals (lines B and C). The melt may be represented as either glass or quench crystals. Glass from liquid must be distinguished from glass formed from dissolved silicate in the fluid. This is done by noting that glass from stable liquid has a higher refractive index and also that glass quenched from fluid often forms minute spheres. Distinguishing quench from stable crystals is an art form with criteria that vary from system to system, but it is almost always possible. Observation of small amounts of enstatite crystals in glass is easy using a grain mount in polarized light. This technique is tedious

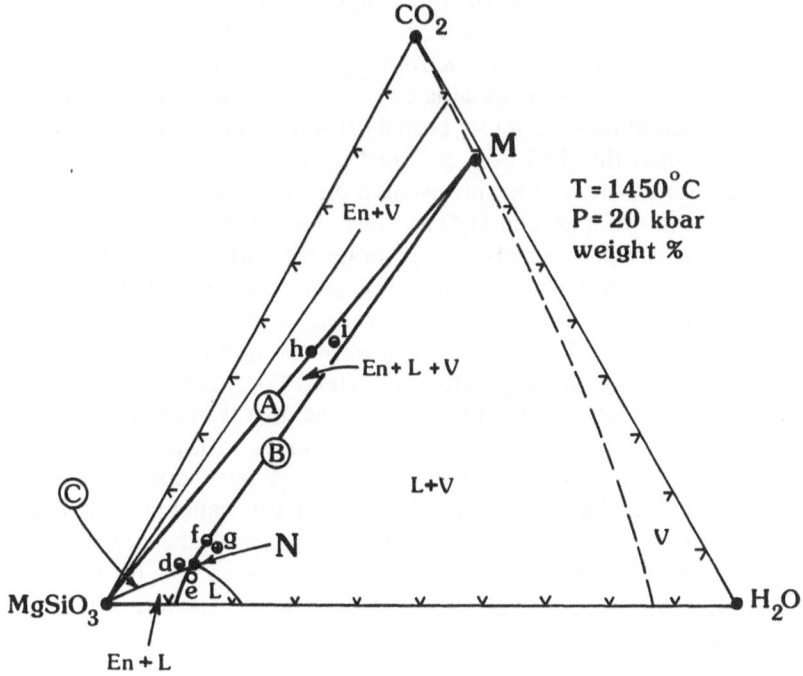

Figure 8.3 Phase equilibria in the system $MgSiO_3-CO_2-H_2O$ at constant P and T. Point N gives the CO_2 and H_2O solubility in fluid and enstatite saturated liquid. See text for discussion. Taken from Eggler (1975b).

because of the large number of experiments required for each $P-T$ point, but it yields highly accurate results and may be necessary to establish the accuracy of other methods, or in cases where quench crystallization prevents the use of direct measurement.

8.3 Spectroscopic measurements and speciation models

Knowledge of the molecular or ionic form that volatiles have in melts makes it easier to formulate thermodynamic models for solubility and also provides insight into the effects that the volatile has on the physical properties of the melt. Energies measured by infrared and Raman spectroscopy are in the range for molecular vibrations and rotations, so these techniques can be used to identify the nature of species such as H_2O, OH^-, CO_2, and CO_3^{2-}. Unfortunately, the modes for these species are very weak in the Raman, so only infrared spectra are really useful for observing such volatiles in silicate melts. On the other hand, Raman spectra are useful in determining the silicate framework structure (McMillan 1984).

Stolper (1982) used infrared spectroscopy to measure molecular H_2O and OH^- in hydrous glasses of basaltic to rhyolitic composition in which the total H_2O content had previously been determined. The analyses were done on disks of glass that had been polished on both sides and in which the thickness was measured with a precision of 0.01 mm. A systematic variation was observed in the molecular H_2O and OH^- contents as a function of total H_2O content. Stolper was able to calibrate the intensity of the infrared absorption bands so that infrared measurements can now be used to analyze silicate glasses for H_2O. In applying the results of structural measurements on glasses to silicate liquids, there is always the question of whether the structure changes on going from liquid to glass. For instance, in the case of H_2O, it is possible that most of the H_2O in the melt is in the form of OH^- and that with falling T one of the protons reacts with an OH^- to form molecular H_2O. A preliminary experiment has been done to test this hypothesis by making a moderate P infrared cell capable of holding a hydrous melt at T up to 850°C. Preliminary results of that experiment show no substantial changes in the infrared spectrum up to 850°C. This result suggests that the molecular H_2O observed in the glass also occurs in the melt (Aines et al. 1983).

Infrared spectra of carbonated glasses have also been measured (Fine & Stolper 1985, 1986). These results show that basaltic and other low-sodium glasses contain only CO_3^{2-} as the dissolved species, but that sodium aluminosilicate glasses contain both CO_3^{2-} and molecular CO_2. The $CO_2 : CO_3^{2-}$ ratio was observed to be a strong function of composition. Fine & Stolper describe their results with the following equilibrium:

$$CO_2 \text{ (molecular, in melt)} + O^{2-} \text{ (melt)} \rightleftharpoons CO_3^{2-} \text{ (melt)}$$

which has the equilibrium constant:

$$K = a_{CO_3^-}\text{-melt}/(a_{O^{2-}} \text{ melt})(a_{CO_2} \text{ melt})$$

and can be rearranged to yield:

$$X_{CO_3^-}/X_{CO_2} = (a_{O^{2-}})(K')$$

This figure should be a constant for a given melt composition providing the activity coefficients of CO_3^{2-} and CO_2 are constant and that the oxide ion activity is constant over the range of total CO_2 concentrations observed. The above relation is consistent with the experimental observations. The fact that different bulk compositions yield different $CO_2 : CO_3^{2-}$ ratios would suggest that the oxide ion activity is a strong function of bulk composition.

138

8.4 Distribution of elements between melts and fluid

Just as crystal/liquid fractionation may change magma composition, so may fluid/melt partitioning cause magmatic fractionation. In order to evaluate the extent to which that process is important, we must know the equilibrium distribution (or partitioning) of elements between coexisting melt and fluid. Unlike crystal/melt partitioning, in which there are strong crystal-chemical controls on the solubility of elements in the crystal, fluid/melt partitioning involves two phases of continuously variable composition. However, because of complex formation in both fluid and melt, some elements may change their partitioning dramatically, depending on the concentrations of other species, such as chloride, in the system.

Some of the goals of fluid/melt partitioning experiments are:

(a) To identify those elements which are significantly partitioned into the fluid. If an element shows essentially no partitioning into the fluid, it will not be affected by fluid fractionation.
(b) For those elements which show significant partitioning into the fluid, to identify the important factors which affect the process.
(c) To measure accurately the partition coefficient under geologically important conditions, for instance by controlling activities of oxygen, H_2O, chloride, and hydrogen ion.

The first example we give was done with the second two goals in mind.

8.4.1   The effect of chloride on rare earth element partitioning

Flynn & Burnham (1978) carried out this study because small differences in rare earth element (REE) patterns may sometimes be used to identify a particular fractionation process. They chose the REE Ce, Eu, Gd, and Yb as representative of all important REE behavior. They sought to determine the partition coefficients, D, for these elements and also to determine the effect of chloride concentration on the partition coefficient. Using Gd as an example:

$$D_{Gd}^{fl/melt} = \frac{\text{Concentration Gd in aqueous fluid}}{\text{Concentration Gd in melt}}$$

Most of their work was done on a natural granite pegmatite composition. All of their experiments were done at 800°C and at pressures of 1.25 and 4 kbar. The REE concentrations used were in the 10 to 100 ppm range. Trace element partitioning experiments are difficult to equilibrate (see Section 7.5), and so Flynn & Burnham (1978) worked out the following procedure to ensure equilibrium:

(a) The sample of Spruce Pine pegmatite was first ground, fused in air, and the quenched glass ground to a fine powder.

139

(b) An aqueous chloride solution was spiked with radioactive isotopes of the chosen REE. The chloride concentrations ranged from 0.5 to 0.9 mol/kg H_2O. The chlorides were added as NaCl, HCl, KCl in the ratios 2 : 2 : 1. This step was necessary to prevent the aqueous fluid from leaching alkalies from the silicate melt during the run.

(c) Equal masses of chloride solution and glass were sealed in capsules and held at a selected P and 800°C for 4 days.

(d) The resulting glass contained most of the REE initially in the chloride solution. However, the distribution of radioisotopes was examined using beta-track autoradiography (Mysen & Seitz 1975) and found to be inhomogeneous.

(e) In order to homogenize the glass, it was ground to <350 mesh and rerun at the same P and T for another 4 or 5 days. The glass resulting from this second cycle was found to be homogeneous.

(f) The glass was crushed and sieved to the 40–60 mesh size range.

(g) The crushed glass was weighed into a small platinum capsule perforated by pin holes. The perforated capsule was placed inside a larger Pt capsule, and an aqueous chloride solution was weighed into the larger capsule which was then welded closed. The mass ratio of glass to chloride solution was 1 : 5 so that there would be a large enough amount of the chloride solution to analyze after the run. The perforated inner capsule is necessary to separate the equilibrium melt from quench material precipitated from the aqueous fluid phase. That quench material is in the form of amorphous powder or glass spheres which cannot be readily distinguished from the glassy quenched melt.

(h) The capsule from step (g) was run at P and T for 4 days.

(i) At the end of the run, the capsule was opened and the aqueous solution rinsed into a container. The capsule was soaked in a solution spiked with a La carrier to prevent adsorption of the REE of interest on the capsule walls. The perforated capsule was cut open and the glass inside removed.

(j) The chloride solution and the glass were analyzed for Ce, Eu, Gd, and Yb using neutron activation analysis. In calculating the REE concentrations to parts per million by weight, the analyses had to be corrected for the decrease in the mass of the glass caused by dissolution in the aqueous fluid, about a 20% to 40% correction.

For each pressure and chloride concentration, as many as 20 samples were run. In some samples the initial ratio of REE in aqueous solution to REE in glass was higher, and in some samples lower, than the equilibrium value. The set of samples thus constitutes a true reversal if both the low and high initial values converge on some intermediate value. The results did show convergence with a two sigma error of ±25%, which is quite good for this kind of experiment.

The results show that for Ce, Gd, and Yb there is a strong linear correlation of D with the cube of the chloride molality. These results suggest reactions of the type illustrated, using Gd as an example.

$$GdCl_3 + 3OH^- \rightleftharpoons Gd(OH)_3 + 3Cl^-$$

fluid   melt   melt   fluid

The reaction is written this way because it is known that chloride itself is partitioned very strongly into the fluid (Kilinc & Burnham, 1972). These experiments demonstrate a strong dependence of D_{REE} on total chloride concentration in the fluid, with D_{REE} changing by a factor of 9-fold over the range 0.5 to 0.9 Cl^- molality. Flynn & Burnham (1978) also found fluid/melt partioning to cause large Eu anomalies and minor REE fractionation.

8.4.2 Chloride partitioning between aqueous fluid and silicate melt

This is an example of an experiment to determine whether a given element, in this case chlorine, strongly prefers one phase or another. The procedure is much more straightforward than the previous example. Kilinc & Burnham (1972) studied the effect of pressure on the partitioning of chlorine between an aqueous chloride solution and granitic composition melts. The same double capsule technique was used with the silicate sample placed in the inner, perforated capsule. At the end of a run, the glass and aqueous fluid were separated and analyzed. The analyses were complicated by the fact that chlorine is lost from the glass sample by volatilization when the glass is dissolved. To overcome this problem, Kilinc & Burnham (1972) mixed a portion of the glass together with lithium tetraborate flux, sealed the mixture in a small platinum capsule, and fused the mixture under pressure. This procedure prevented loss of chlorine.

The value of the partition coefficient $D_u^{fl/melt}$ ranged from 53 at 2 kbar, to 50 at 6 kbar, to 13 at 8 kbar. Thus at pressures below 6 kbar, almost all of the chlorine is in the fluid. There is an intriguing decrease in D from 6 to 8 kbar, the highest pressure investigated. It would be interesting to do an experiment at 10 kbar.

In experiments such as these with very large D values, high precision determinations are probably unnecessary. Fairly good estimates can be made by analyzing for the element in question in the phase in which it has the lowest concentration, in this instance the glass, and calculating the partition coefficient by mass balance.

141

8.5 Conclusions

The intent of this chapter was to present a brief overview of some of the most used experimental techniques for the study of interactions between silicate melts and either aqueous or carbonic fluids (or both). We have only covered a small proportion of the many approaches used and, because this is an active area of research development, new techniques and experiments may be expected in the near future.

9 Physical properties of Earth materials

9.1 Introduction

Knowledge of the physical properties of minerals and silicate melts is important because the Earth is dynamic, not static, and most of the geologic processes that occur in it are the results of motion. The movement causes disruption of equilibrium. For instance, density differences cause magma to rise so that both the magma and its country rock encounter new $P-T$ regimes. The rate and mechanism of rise depends on many physical properties, including the density and viscosity of the magma and of the country rock. The rates at which the magma crystallizes and the country rocks react control how closely equilibrium is approached, and determine the rate at which heat is consumed or released by the process. Thus in order to develop large scale models for Earth processes, we must know the densities, viscosities, diffusion coefficients, crystal growth rates, and elastic properties of crystals and melts.

9.2 Viscosity of silicate melts

Magmas may consist entirely of silicate melt or mixtures of melt, crystals, and vapor bubbles. The viscosity relations in the mixtures may be very complex and will not be considered here. The viscosity of naturally occurring silicate melts ranges over about 13 orders of magnitude. Melt viscosity at atmospheric pressure is well known, mainly from studies by glass scientists and by Georges Urbain (see the summary by Urbain *et al.* 1982). The atmospheric pressure research demonstrates the very strong effects of bulk composition and temperature on viscosity. Much less is known of the effects of pressure and H_2O content on melt viscosity, but elegant experiments to measure those effects have been done and will be described here.

The first high pressure experiments were done by Shaw (1963b) to measure the effect of H_2O on the viscosity of a granitic bulk composition. Shaw's experiments were done in cold-seal pressure vessels at temperatures just above the H_2O-saturated liquidus. Kushiro (1976) adapted Shaw's technique to the piston-cylinder furnace assembly which allowed investigations at much higher $P-T$ conditions (up to 1500°C and 25 kbar). The method involves placing spheres of a material more dense than the silicate melt at the top of a capsule and measuring the distance the spheres sink in a given time. The sinking distance, d, is given by the Stokes equation:

$$d = \frac{2a^2 g \Delta \rho t}{9\eta}$$

where a is the radius of the sphere, g the gravitational acceleration, $\Delta\rho$ the difference in density between melt and sphere, t the elapsed time of the experiment and η the viscosity of the melt. The spheres may be made of metals such as platinum, gold, or silver, or they may be a mineral which is in equilibrium with the melt at the $P-T$ conditions of the experiment. If different density spheres, or spheres with different sizes are used, then the density of the melt may be determined as well as the viscosity. The simplicity of the method is illustrated in Figure 9.1 in which results from a series of experiments done by Kushiro (1976) are shown. For runs of the same T, elapsed time and sphere diameter, simple visual inspection shows the effect of P on melt viscosity. Kushiro's results for a melt of jadeite composition are shown graphically in Figure 9.2.

This technique assumes that the viscosities of the silicate melts are Newtonian, that is, that the viscosity is independent of strain rate. This

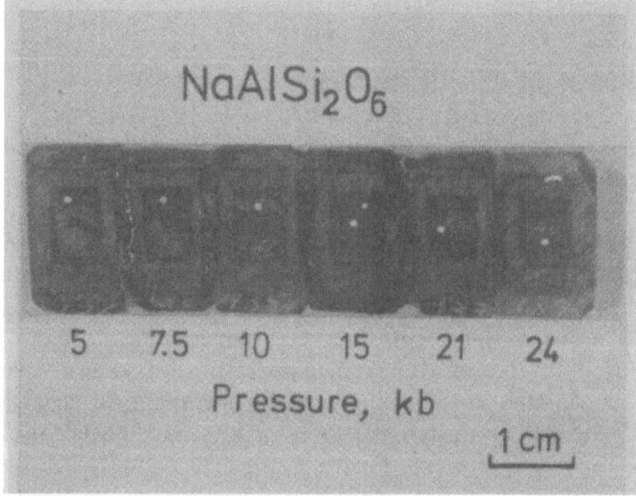

Figure 9.1 Photographs of cross-sections of samples from Kushiro's experiments. The viscosity is inversely proportional to the distance the spheres (bright spots) have fallen. Run duration was 8 min except for the 5 kbar run which was 10 min. Temperature was 1,350°C for each experiment. (From Kushiro 1976).

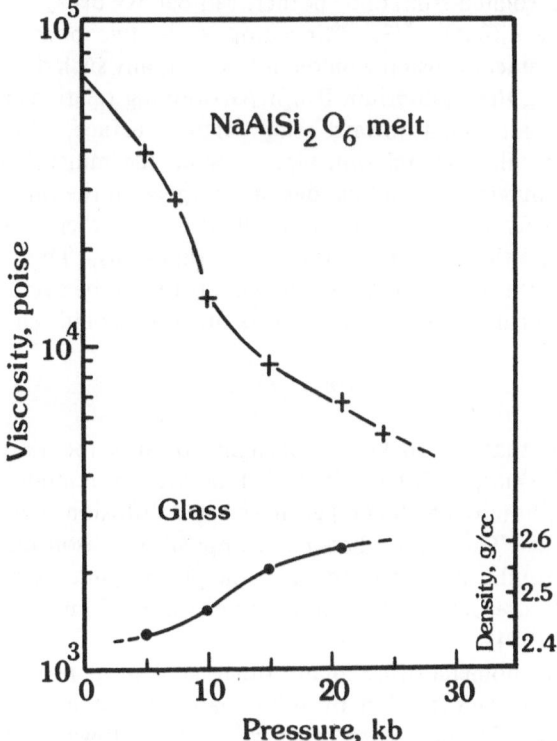

Figure 9.2 Viscosity of melts of NaAlSi$_2$O$_6$ composition as a function of P at 1,350°C. Calculated using the experiments shown in Figure 9.1. Measured densities of quenched glasses shown at bottom right.

assumption has not been rigorously tested, but it could be investigated by using spheres of the same diameter but greatly different density.

9.3 Diffusion

Diffusion is an extremely important petrological phenomenon. If a rock were to crystallize under equilibrium conditions of P, T, a_{H_2O} etc. and remain forever at those conditions, then all of the minerals in the rock would be chemically homogeneous and would remain at constant composition. Diffusion, which is the random movement of atoms through the crystal (or liquid) structure, would still be occurring, but it would not be easy to detect because the flux of atoms through the boundary between two phases would be the same in both directions. It could however affect isotopic ratios in the case in which, for example, two coexisting minerals have different equilibrium Rb and Sr concentrations. If the minerals were isolated from one another, radioactive decay of ^{87}Rb to ^{87}Sr would produce different ^{87}Sr : ^{86}Sr ratios in the two phases, but if they remained in contact diffusion would tend to erase any such differences while still maintaining the equilibrium Rb/Sr partitioning relationship.

As a rock is brought towards the Earth's surface, $P-T$ conditions change, and equilibrium relationships between the minerals change. The ability of the minerals to readjust depends largely on the rates of diffusion through the different mineral, melt and fluid phases, their physical dimensions, and the time spent at the new conditions. This statement is summarized in the approximate relationship between the average distance x which an atom can diffuse in time t if its diffusion coefficient is D;

$$x \approx \sqrt{D \cdot t}$$

It is apparent that, as rocks are brought towards the Earth's surface, chemical and isotopic distributions will be frozen-in under a range of different conditions depending on grain size and diffusion coefficients of the different atoms. Thus, the chemical compositions, isotopic ratios, and extent of disequilibrium provide, in principle, considerable information about the pressure, temperature and time histories of mineral assemblages (e.g., Lasaga 1984).

Diffusion phenomena affect many other geological processes. At high temperatures the deformation of solid material is controlled by atomic diffusion. Thus, the mechanical properties of the lower crust and upper mantle with all their tectonic implications should be considered in terms of rates of atomic migration through crystal lattices. In natural silicate melts, the establishment and eradication of chemical gradients are controlled by the diffusivities of migrating species. For example, the rates of assimilation of wall rock or xenoliths by ascending magma depend, in part, on diffusion away from the melt/xenolith or melt/wall-rock interface.

There are a number of different usages of the term "diffusion coefficient," so in an attempt to avoid confusion, we will begin by defining them. A tracer diffusion experiment measures D^*, the diffusivity (cm^2 . s^{-1} or m^2 . s^{-1}) of a component at trace concentration levels (< 100 ppm) in a matrix of constant composition. Thus one might, for example, determine the tracer diffusion coefficient for Mg diffusion in pure Fe_2SiO_4 fayalite. If the trace species is chemically the same as the host, but can be distinguished from it (e.g., a radioactive isotope of Fe in nonradioactive Fe_2SiO_4), then a "self-diffusion" coefficient is obtained. An "interdiffusion coefficient" ($\bar{D}$) is obtained from an experiment where there is diffusive exchange of two or more species across a boundary in a system which exhibits a true chemical compositional or chemical potential gradient (Fig. 9.3). Such interdiffusion coefficients are generally composition-dependent.

In practice, diffusion coefficients are measured either by determining the

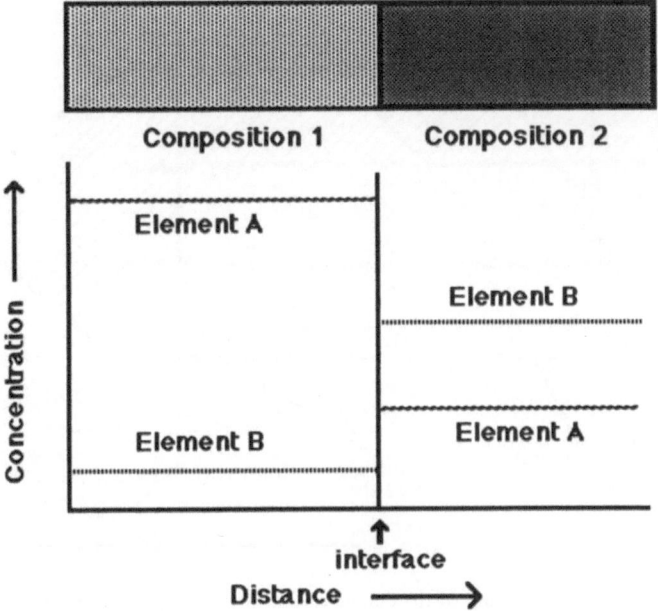

Figure 9.3 Schematic diagram showing two samples in contact. Side 1 has different concentrations of elements A and B than does side 2. The concentration differences result in a chemical potential gradient across the interface that causes diffusion to occur. This is an example of an experimental set-up for an interdiffusion experiment.

rate of material transport across an interface or, more commonly, by measuring the concentration profile of the diffusing species after known time.

Watson (1979a) measured the tracer diffusion of ^{45}Ca in sodium–calcium–aluminosilicate melt at high pressure. He used an experimental arrangement (Fig. 9.4) in which the tracer was deposited on the end of a slug of glass before sealing it in a capsule and bringing it to P and T. Fick's Second Law of Diffusion relates the rate of change of concentration with time at some point in the medium to the second derivative of concentration with respect to distance x:

$$\frac{\partial C}{\partial t} = D \cdot \frac{\partial^2 C}{\partial x^2}$$

A solution to this equation for the geometry used by Watson with a tracer source diffusing into a semi-infinite medium is (e.g., Crank 1975):

$$\ln C = A - \frac{x^2}{4Dt} \tag{9.1}$$

147

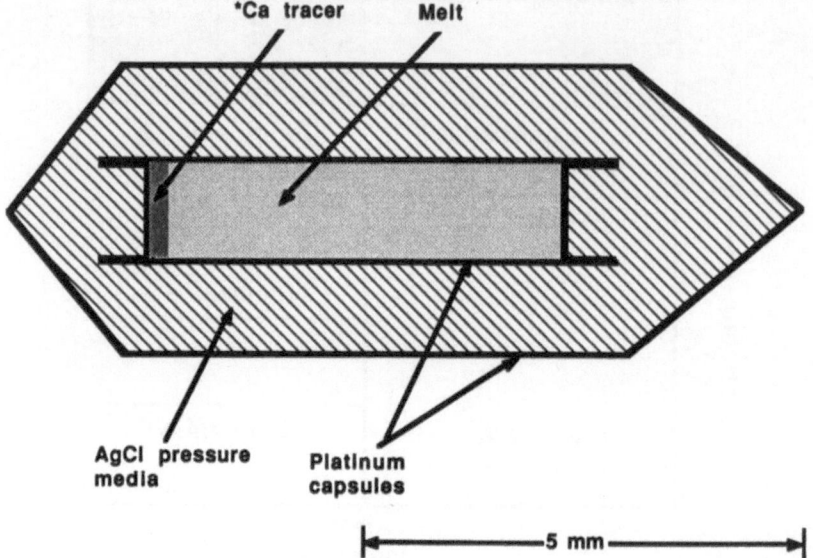

Figure 9.4 Cross-sectional view of the double capsule arrangement used by Watson (1979a) for tracer diffusion experiments. The AgCl in the outer capsule provides a hydrostatic pressure medium to prevent deformation of the sample.

In Equation (9.1) A is a constant which has to be determined from the experimental data and C is the concentration at distance x after time t. Watson determined relative concentrations of the tracer isotope by making lengthwise sections of the capsule and recording the activity of the β particle emitting ^{45}Ca on nuclear emulsions (autoradiography). A scan of β-track density in the direction of diffusion gave relative concentration which enabled the diffusion coefficient to be obtained by plotting $\ln C$ versus x^2 (Fig. 9.5). This method was used to obtain tracer diffusion coefficients for the $P-T$ range 1 to 30 kbar and $1,100-1,400°C$.

Diffusion is a process in which an atom jumps from a "comfortable" position in the structure of the host to a comparable neighboring position. In making the jump the atom has to pass between other relatively closely packed atoms, and it momentarily enters a hostile environment *en route* to its new position. The intermediate position is an unstable, high-energy configuration, and a diffusing atom must have enough energy to clear this barrier before it reaches the neighboring stable position. If the energy barrier is of height E_a (cal . mol^{-1} or J . mol^{-1}), then the number of atoms having sufficient energy to cross it increases with increasing temperature because of higher thermal energy. The diffusion coefficient D increases with temperature, therefore, in a manner described by the Arrhenius relation-

148

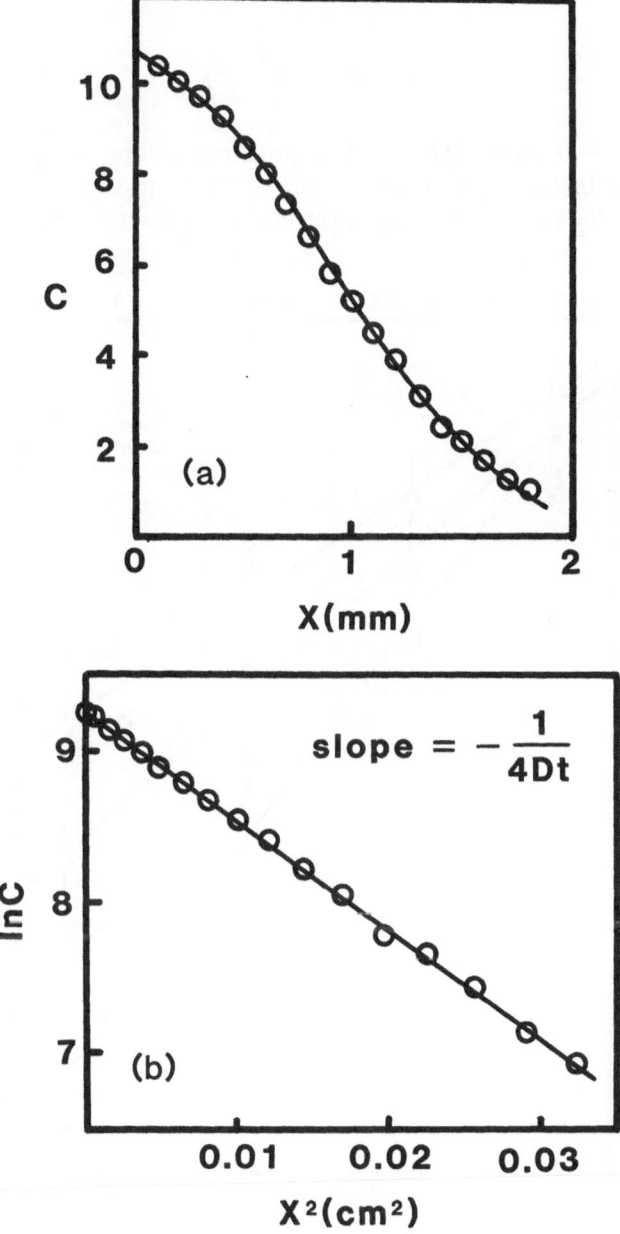

Figure 9.5 Diffusion profile for a run made at 30 kbar, 1,300°C. C is a composition parameter of arbitrary units, proportional to the activity of ^{45}Ca in the melt; X is a distance from the thin source. A linearized diffusion profile with least-squares fit to the data points (Watson 1979a) is shown in the lower figure.

ship:

$$D = D_0 \exp\left(\frac{-E_a}{RT}\right) \qquad (9.2)$$

A plot of $\ln D$ versus $1/T$ (Fig. 9.6) produces a slope of $-E_a/R$ and an intercept of $\ln D_0$. The latter, although not exactly predictable for complex phases, is related to the entropy of the intermediate, activated state. As can

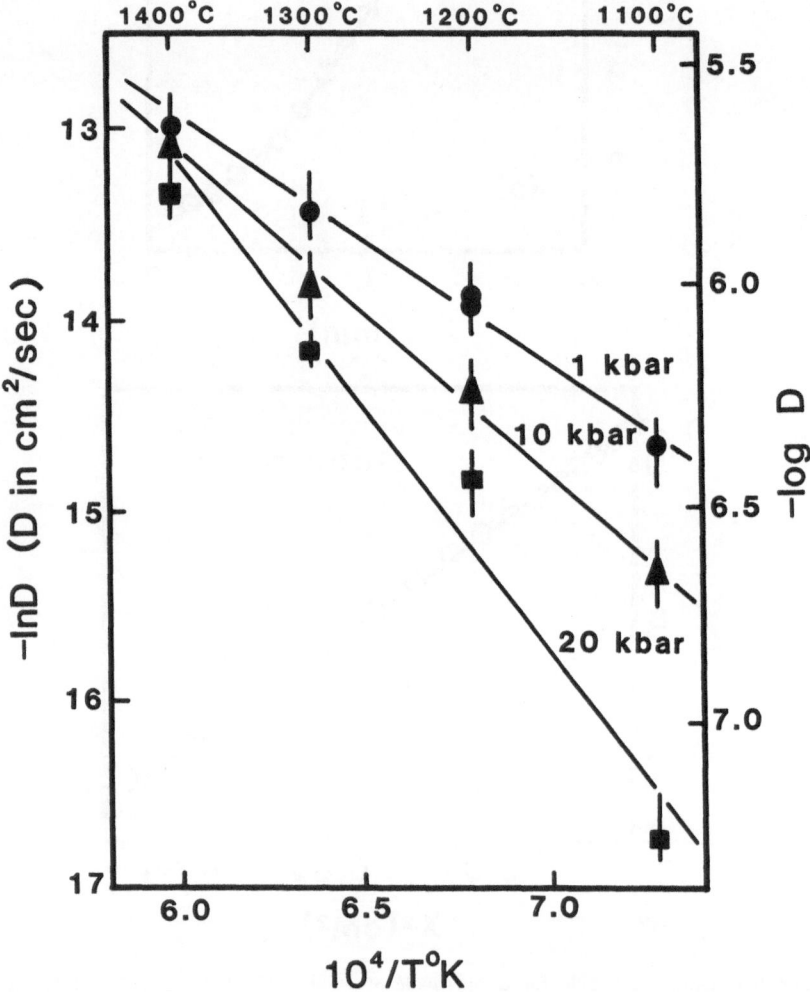

Figure 9.6 Natural logarithm of the diffusion coefficient (D) of calcium plotted against reciprocal absolute temperature. Lines are least-squares fits to three isobaric series of runs [1, 10, and 20 kbar; see Watson (1979a)].

be seen from Figure 9.6, diffusion coefficients depend on pressure as well as temperature because production of the intermediate state generally requires a volume change as well as an activation energy. The magnitude of this volume change is impossible to predict in complex phases, and it is only known for a few experimental systems. Generally the effect of pressure is regarded as subordinate to that of temperature. Watson's experiments do, however, show a pressure dependence of D, with an "activation volume" that is much smaller at high temperatures than at low temperatures (Fig. 9.6). He explained this observation by suggesting that, at high temperatures (1,400°C in this case), the melt has much more void space than at low temperature and that, under such conditions, the intermediate step requires virtually no displacement of neighboring atoms by the migrating species.

Most studies of melt and glass diffusivity (e.g., Margaritz & Hofmann 1978) have used similar radiotracer techniques to those described here. The recent development of high precision ion microprobes has, however, greatly expanded the range of diffusion studies which can be performed. Karsten et $al.$ (1982) used this instrument to determine the concentration profile of hydrogen (attached to H_2O molecules) diffusing into an obsidian glass at $PH_2O = 700$ bar and temperatures from 650 to 950°C. Their analyses of the concentration profiles showed that the H_2O interdiffusion coefficient is strongly dependent on H_2O content in the concentration range 0.2 to 4 wt% H_2O. They also showed that H_2O diffusion was not very temperature dependent, with an activation energy of 79.5 kJ/mol. Lapham et $al.$ (1984) used similar techniques to show that there is no measurable pressure effect on H_2O diffusion in this system at pressures between 700 bar and 5 kbar.

Because the building and testing of models is the major goal of experimental science, it should be pointed out that there has been a very successful empirical law applied to diffusion in silicate melts. This is the compensation law of Winchell (1969) which asserts that there is a simple relationship between D_0 and E_a for many species diffusing in a host of constant bulk composition (i.e., tracer diffusion). The "compensation" law is given by

$$E_a = a + b \log D_0 \qquad (9.3)$$

and a plot of E_a against $\log D_0$ (Fig. 9.7) shows that it works very well for alkali-silicate glasses.

In the context of natural systems, Hart (1981) has shown that the compensation relationship works well for basalt liquid and obsidian and also for crystalline olivine and feldspar. The implication of the compensation model is that, for a particular host, there is one temperature at which most species have the same tracer diffusion coefficient. A good idea of the tracer diffusion coefficient D^*, for an unstudied species, may therefore be obtained by a measurement at one temperature combined with the

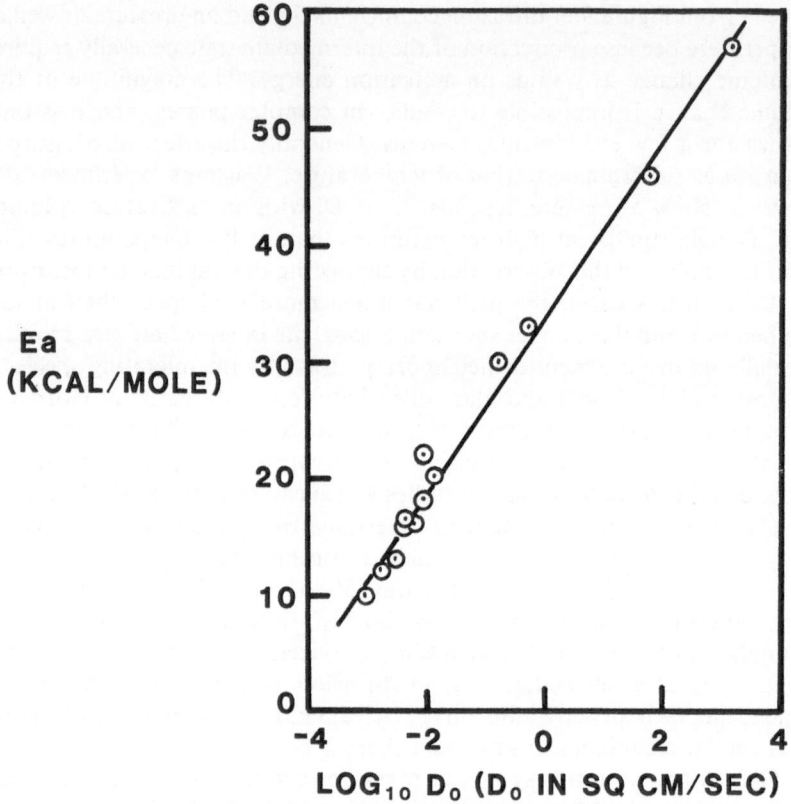

Figure 9.7 Example of compensated diffusion in silicates for the transport of Na, K, and Ca in $Na_2O/K_2O-R_xO-4SiO_2$ (R = Li, Na, K, Mg, Sr, Ca, Ba) glasses (from Winchell 1969).

"common" value of D^* and T. The two numbers, one measured and one model, give the approximate temperature dependence of D^*.

Diffusion in crystalline materials may be investigated with techniques similar to those outlined above for melts and glasses. The diffusive process is, however, rather better understood in crystals than it is in melts. This is because the former have well defined lattice geometries and lattice positions on which the atoms reside, whereas liquids have looser structures with wider varieties of environment of migrating species. In crystals at equilibrium, the numbers of lattice vacancies increase with increasing temperature. Atomic migration through the crystals generally takes place into these vacancies. Since atomic diffusion depends on the presence or absence of vacancies, diffusion coefficients are proportional to the fraction of site vacancies, $[V]$ available for migrating atoms to occupy. Diffusion coefficients also increase at constant vacancy concentration due to the Arrhenius relationship. Therefore, the following dependence of D on $[V]$ and T and is often

152

observed:

$$D \propto [V] \exp(-E_a/RT) \tag{9.4}$$

The concentration of vacancies depends on temperature and on the presence or absence of impurities in the crystal. Consider, for example, the formation of vacancies in pure fayalite, Fe_2SiO_4, coexisting with quartz and, as a simplification, assume that these are dominantly in the Fe^{+2} (Me) sublattice and the O^{2-} sublattice. The equilibrium describing formation of the vacancies may be written:

$$1/2SiO_2 + Fe^x_{me} + O^x_0 \rightleftharpoons V''_{me} + V^{\cdot\cdot}_0 + 1/2Fe_2SiO_4 \text{ (surface)} \tag{9.5}$$

where the Fe^x_{me} and O^x_0 species refer to atoms on their normal lattice positions and V''_{me} and $V^{\cdot\cdot}_0$ to vacancies on the two sublattices. In terms of atomic fractions of defects $[V''_{me}]$ and $[V^{\cdot\cdot}_0]$ on the Fe^{2+} and oxygen sublattices, the equilibrium constant is given approximately by:

$$K_5 = \frac{[V''_{me}][V^{\cdot\cdot}_0] a_{Fe_2SiO_4}^{ol \, 1/2}}{[Fe^x_{me}][O^x_0] a_{SiO_2}^{1/2}} \approx [V''_{me}] \cdot [V^{\cdot\cdot}_0] \approx \exp\left(\frac{-\Delta G_5^0}{RT}\right)$$

where activity coefficients have been ignored and the site fractions of the dominant species $[Fe^x_{me}]$ and $[O^x_0]$ are assumed to be approximately 1.0. Activities of SiO_2 and Fe_2SiO_4 are also 1.0 in the fayalite–quartz assemblage. The superscripts on the vacancy concentrations, V''_{me} and $V^{\cdot\cdot}_0$ refer to the fact that an Fe^{2+} vacancy has a charge of -2 and an oxygen vacancy a charge of $+2$.

Consider now what would happen if there were small amounts of ferric iron occupying Fe^{2+} positions and tetrahedral Si^{4+} positions. The former defect Fe^{3+}_{me} has net charge $+1$, denoted $Fe^{\cdot}_{me}$ and the latter Fe^{3+}_{si} charge -1 denoted Fe'_{si}. Applying the constraint of charge balance in the crystal as a whole the total number of the different defects must sum up as follows:

$$2V''_{me} + Fe'_{si} = Fe^{\cdot}_{me} + 2V^{\cdot\cdot}_0$$

In general, the equilibrium constants for equilibria such as (9.5) increase rapidly with increasing temperature. This means that the product $[V''_{me}] \cdot [V^{\cdot\cdot}_0]$ tends to be very small at low temperatures. Hence the defect concentrations in fayalite at low temperatures have to depend primarily on how much Fe^{3+} has been introduced into the structure as $Fe^{\cdot}_{me}$ and Fe'_{si}. That is, the ferric iron impurity generally controls the production of defects at low temperatures. If $Fe^{\cdot}_{me}$ dominate over Fe'_{si}, the charge balance constraint simplifies to:

$$Fe_{me}^{\cdot} \approx 2V_{Fe}''$$

If Fe_{Si}' are more important the relationship is:

$$Fe_{Si}' \approx 2V_0^{\cdot}$$

As temperature increases, the product $[V_{me}'] \cdot [V_0^{\cdot}]$ becomes larger and at some high temperature these intrinsic defects dominate over the ferric impurity defects, i.e.,

$$V_{me}'' = V_0^{\cdot} \text{ (high temperature limit)}$$

In this regime, the diffusion coefficients of Fe and O in fayalite depend solely on the intrinsic vacancy properties of the mineral since impurities do not contribute markedly to vacancy concentration. At low temperatures the vacancy concentrations and diffusion coefficients D depend substantially on the amounts of ferric impurity present and also on which type of Fe^{3+} ($Fe_{me}^{\cdot}$ or Fe_{Si}') predominates. This is known as the "extrinsic region." A plot of how D_{Fe} might look as a function of temperature in intrinsic and extrinsic regimes is shown in Figure 9.8.

Nakamura & Schmalzried (1983) studied the defect-controlled stoichiometry of olivine solid solutions (fayalite$_{100}$ to fayalite$_{20}$forsterite$_{80}$ compositions) in the temperature range 1,000 to 1,280°C at 1 atm pressure. They did this by careful measurement of the weight changes observed in large (1–2 g) pellets of synthetic olivine equilibrated inside a thermobalance with a CO_2/CO atmosphere of known oxygen fugacity. In the absence of other phases, the observed weight changes refer entirely to the addition of oxygen to, or removal of oxygen from, the olivine with corresponding production of Fe^{3+}, Fe^{2+}, and vacancies. In treating their data, Nakamura & Schmalzried assumed that, in the extrinsic region, V_{me}'', Fe_{Si}', and $Fe_{me}^{\cdot}$ are the majority defects in the structure. Given this assumption, the reaction for incorporation of oxygen into olivine may be written as:

$$10Fe_{me}^x + Si_{Si}^x + 2O_2 \rightleftharpoons 3V_{me}'' + 7Fe_{me}^{\cdot} + Fe_{Si}' + Fe_2SiO_4 \qquad (9.6)$$

For which the equilibrium constant (K_6) may be approximately expressed in terms of defect site fractions, $[V_{me}'']$ etc., as

$$K = \frac{[V_{me}'']^3 \cdot [Fe_{me}^{\cdot}]^7 \cdot [Fe_{Si}'] \cdot a_{Fe_2SiO_4}}{[Fe_{me}^x]^{10} \cdot [Si_{Si}^x] \cdot fO_2^2}$$

Noting as before that $[Fe_{me}^x]$ and $[Si_{Si}^x]$ are essentially 1.0 and that $a_{Fe_2SiO_4}$ is

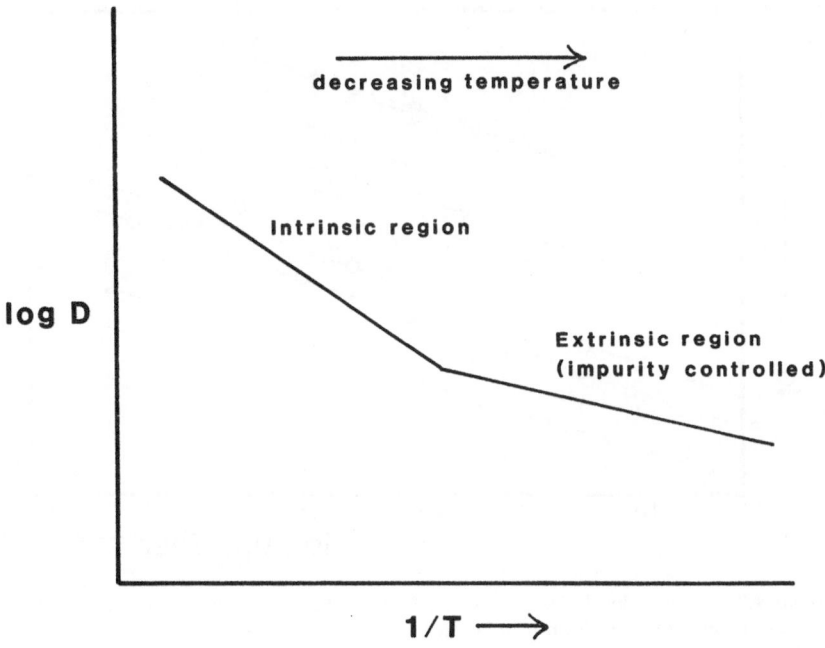

Figure 9.8 Schematic showing expected T dependence of cation diffusion in silicates.

fixed at constant olivine composition gives:

$$[V''_{me}]^3 \alpha \frac{fO_2^2}{[Fe^{\cdot}_{me}]^7[Fe'_{Si}]}$$

Since, however, no other phases were produced or consumed during the experiments, the stoichiometry of Reaction (9.6) forces a proportionality between V''_{me}, Fe'_{Si}, and $Fe^{\cdot}_{me}$

$$7Fe'_{Si} = Fe^{\cdot}_{me} = 7/3 V''_{me}$$

which gives:

$$[V''_{me}]^{11} \approx PO_2^2$$

$$[V''_{me}] \approx PO_2^{1/5.5}$$

Figure 9.9 shows a plot of nonstoichiometry δ (proportional to V''_{me} above) versus log PO_2 for the olivine solid solution series investigated by Nakamura & Schmalzried. Close to the forsterite end of the series, vacancy concentration obeys the $PO_2^{1/5.5}$ relationship discussed above. At higher mole fractions of Fe_2SiO_4 however, δ approaches a one-fifth power

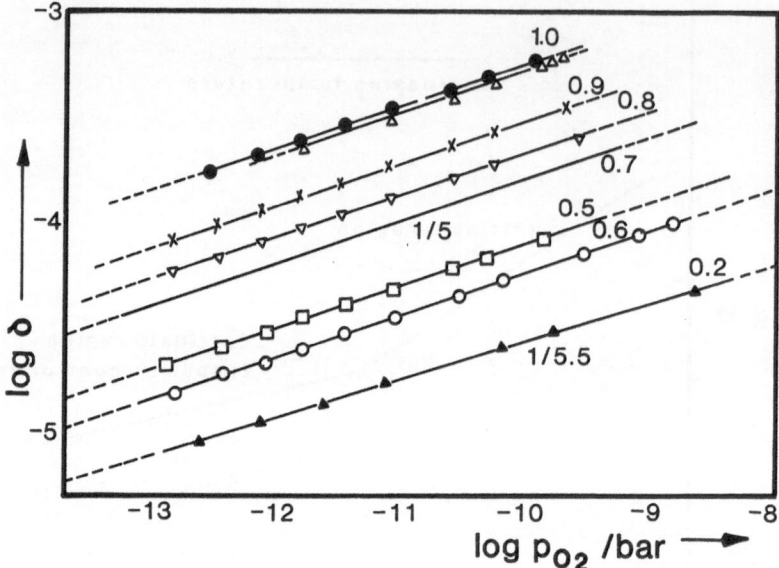

Figure 9.9 Log δ vs. log P_{O_2}-plots at $T = 1,130°C$ for Fe–Mg olivine with the indicated mole fraction of Fe_2SiO_4 (1.0 down to 0.2). Slopes of curves are between 1/5 and 1/5.5. (from Nakamura and Schmalzried 1983).

dependence on PO_2. Nakamura & Schmalzried interpreted this observation as being due to increasing association of Fe'_{Si} and Fe'_{me} forming a charge-balanced $(Fe'_{Si}Fe'_{me})$ species at high concentrations of Fe in the system. It may readily be shown, by the same arguments as those above, that complete association of Fe'_{me} and Fe'_{Si} in Reaction (9.6) leads to a $PO_2^{1/5}$ dependence of $[V'''_{me}]$. Having obtained relationships between defect concentration and oxygen fugacity, it is now appropriate to consider whether or not cation diffusion in olivine really does have the vacancy dependence that was suggested earlier.

Buening & Buseck (1973) determined interdiffusion coefficients for Fe^{2+}–Mg diffusion in olivine. This was done by taking oriented single crystals of natural olivine of Fe:Fe + Mg ratio of 0.08 and making diffusion couples (Fig. 9.3) with powdered synthetic fayalite. After cutting faces perpendicular to the three olivine crystallographic axes, the large natural crystal had fayalite packed around it, and the couple was annealed in a furnace for 192–336 hours at temperatures of 1,000–1,200°C. Oxygen partial pressure was controlled in the range 10^{-12} to 10^{-14} bar by using a CO/CO_2 atmosphere of known, fixed composition. After quenching in argon, concentration–depth profiles were determined along each of the three crystallographic axes of the single crystal, using an electron microprobe.

The nature of the diffusion couple, one side fayalite and the other side almost pure forsterite, allows the determination of interdiffusion coefficients $\bar{D}$ for several compositions between the two ends of the couple. Some of Buening & Buseck's data are shown in Figure 9.10. It can be seen that, at $1,100°C$, diffusion along the C axis is dependent on the $Fe : Fe + Mg$ ratio of olivine. The interdiffusion coefficient $\bar{D}_{FeMg}$ at fayalite$_{65}$ is about seven times greater than at fayalite$_{13}$. It also depends on P_{O_2}. From Equation 9.4 it is predicted that $\bar{D}_{FeMg}$ should be proportional to $[V''_{me}]$. Since Nakamura & Schmalzried's data indicate that $[V''_{me}]$ depends on P_{O_2} in the range between 1/5 and 1/5.5, the diffusion coefficients should be affected by P_{O_2} in a similar manner.

Figure 9.10 shows a comparison of slopes for $P_{O_2}^{1/6}$ and $P_{O_2}^{1/5}$ dependence of D_{FeMg} with the observed effects of P_{O_2} on $\bar{D}_{FeMg}$. The agreement is very good. In fact, from a complete analysis of their data, Buening & Buseck concluded that the dependence of D on P_{O_2} in this temperature–composition range was between 1/6.7 and 1/5.1. Their observations are in very good agreement with Nakamura & Schmalzried's vacancy model and confirm the influence of vacancies on cation diffusion in olivine.

In conclusion, it appears that diffusion through crystal lattices is vacancy-

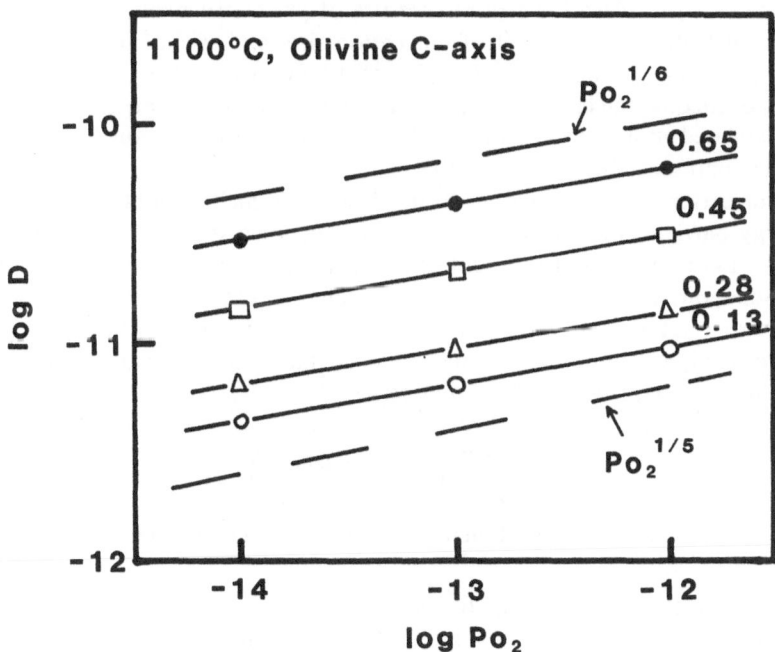

Figure 9.10 Interdiffusion coefficient along olivine C-axis from Buening and Buseck (1973). Curves labeled with Mol-fraction fayalite. Theoretical curves for P_{O_2} dependence shown for comparison.

dependent and that in many geological situations the vacancy concentrations are controlled by oxygen fugacity. It is necessary, therefore, to ensure that appropriate f_{O_2} constraints are applied when measuring diffusion coefficients or attempting to apply them to the interpretations of geological processes (even when the diffusing species is present in only one oxidation state).

9.4   Crystal growth

The application of most experimental data to rocks requires the assumption of chemical equilibrium in nature. Disequilibrium has also been discussed from the standpoint of diffusion during reequilibration. The other important requirements for the maintenance of equilibrium are that the stable phases nucleate as soon as their stability fields are reached, and that they grow such that the system is just saturated in them. In practice, a small amount of overstepping of equilibrium is required to produce new nuclei, and these nuclei can grow only if $\Delta G_r < 0$, i.e., if there is a small free energy driving force. A number of studies have been aimed at determining the extents of overstepping required for nucleation and for crystal growth. In the case of nucleation from a mixture of pre-existing solids (heterogeneous nucleation) the number of nucleation sites and ease of nucleation are very difficult to model, and will not be discussed any further here. Homogeneous nucleation from single-phase fluids or melts has been investigated more successfully, and the succeeding discussion follows the review paper of Kirkpatrick (1981).

Nucleation of crystals from silicate glasses and melts is controlled by two energy barriers: the free energy of formation of a stable nucleus and the activation energy for attachment of atoms to the growing nuclei. The latter is analogous to the barrier that atoms have to cross in order to diffuse through crystal lattices. The nucleation rate I (nuclei/cm^3) has a slightly more complex temperature dependence than the diffusion coefficient because of the presence of two barriers:

$$I \propto \quad \exp\left(\frac{-\Delta G_a}{RT}\right) \cdot \exp\left(\frac{-\Delta G_{nucl}}{RT}\right) \tag{9.7}$$

barrier to attachment   barrier to nucleus formation

The free energy of nucleus formation is large at high temperatures and becomes smaller with decreasing temperature as the free energy difference between crystal and liquid increases. At low temperatures the barrier to attachment is large, but it becomes smaller with increasing temperature in a way analogous to the diffusion case discussed previously. The interaction of

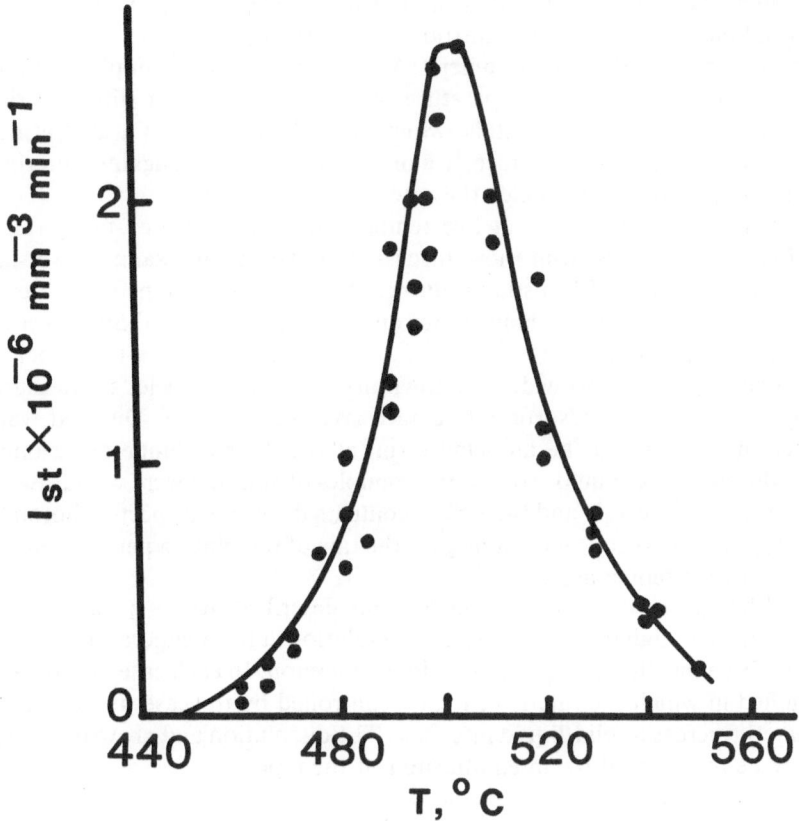

Figure 9.11 Steady state nucleation rate for the composition $2Na_2O-CaO-3SiO_2$ versus temperature. Attachment kinetics dominate at low T and nucleus formation at high T (after Kalinina *et al*, 1980 and Kirkpatrick 1981).

these two terms results in there being a peak on the nucleation rate–temperature curve (Fig. 9.11).

Experimentally, studies of the nucleation rates of silicate liquids are made by melting the composition of interest and rapidly quenching it to a glass. The glass is then run at the experimental temperature to produce a certain number of nuclei in known time. Since the rate of growth of nuclei is generally very small at temperatures where they form in abundance, the nuclei cannot be seen. The nucleation step is therefore followed by a high-temperature anneal in order to grow the nuclei to a reasonable size. The run is then quenched and any heterogeneous nuclei, formed on the capsule wall or at the melt/atmosphere interface, are removed by grinding. The charge is sectioned and the number of nuclei, now visible, counted.

Crystal growth from melts onto pre-existing nuclei is studied by heating the melt above the liquidus, then cooling it to the desired temperature above the solidus. A seed crystal is introduced and the rate of growth of the seed photographed with a movie camera. Alternatively, several samples of liquid are heated at the same temperature above the solidus for different time periods. The lengths of crystals, which generally nucleate at a melt/atmosphere or melt/capsule interface, are measured and plotted against run time. The slope of the graph yields the rate of crystal growth.

Dissolution and growth at low temperature is investigated using rather different techniques from those discussed above. As an example, we shall consider the work of Rimstidt & Barnes (1980) who determined the rates of dissolution into water of amorphous and crystalline forms of SiO_2 over the temperature range of 18 to 305°C. They used a Barnes-type rocking autoclave (Ch. 3) into which a known mass of solid particles of measured (by gas monolayer adsorption) surface area were packed. Distilled water was introduced and the autoclaves run at the desired temperature under conditions of fixed fluid : solid ratio. Samples of solution were withdrawn at appropriate intervals and their SiO_2 contents determined spectrophotometrically. The SiO_2 concentrations gave the rate of dissolution per unit surface area at each temperature.

Although the two types of experiments described here − crystal growth from melt at high temperatures, and dissolution at low temperatures − seem very different, they actually have a lot in common. In each case, a process is studied in which the rate is generally controlled by the ease of transfer of material across a solid/liquid interface. The dissolution and growth reaction may be represented by an equilibrium of the type:

$$A \; \rightleftharpoons \; A$$

solid   liquid

with rates of forward (dissolution) and reverse (growth) reactions being given by, respectively:

$$R_d = K_d a_A^s$$

$$R_g = K_g a_A^l$$

In these expressions, a_A^l and a_A^s are the activities of A in liquid and solid phases and K_d, K_g are rate constants for the forward and reverse reactions. Note that the high-temperature crystallization experiments are aimed at measuring R_g and the low-temperature dissolution experiments R_d, but that both dissolution and growth processes occur simultaneously in all such experiments. Thus, what is actually determined is R_{net}, the difference between the rates of dissolution and growth under the conditions of the experiment. From a combination of the equilibrium constant $K = (K_d/K_g)$

and transition state theory (e.g., Lasaga 1981) it may be shown that R_{net} is given by:

$$R_{net} = R_d - R_g = K_d\left[1 - \exp\left(\frac{\Delta G_r}{RT}\right)\right] \qquad (9.8)$$

where ΔG_r is the free energy driving force of the reaction. Thus, provided some other process, such as diffusion through the melt, is not the rate-limiting step, the same rate relationship applies to both growth and dissolution (Kirkpatrick 1981). Furthermore, Equation 9.8 may be simplified for near-equilibrium conditions ($|\Delta G_r| \ll RT$) to:

$$R_{net} = -K_d \frac{\Delta G_r}{RT} \qquad (9.9)$$

This relation means that, close to equilibrium, the net rate of reaction is a linear function of temperature since ΔG_r is a linear function of temperature. In addition, for an equal temperature difference in the opposite direction, ΔG has an equal value but opposite sign, so that the net rate of dissolution will be equal ($\Delta T = +$) to the net rate of growth ($\Delta T = -$).

Equation 9.8 was used to extract values of K_d for quartz from Rimstidt & Barnes' experiments (Figure 9.12). As may be seen from the ln K versus $1/T$ plot of Figure 9.12, the dissolution data obey the same kind of Arrhenius relationship as do diffusion coefficients. This is to be expected for any process which has an activation barrier over which atoms must pass as they detach from the mineral surface. In this case, the activation barrier involves breaking the Si—O bond in quartz in order to form the aqueous $SiO_2 . nH_2O$ complex in solution. The energy barrier appears to be about 75 kJ mol^{-1}.

The rates of surface attachment and detachment are extremely important in controlling reactions that involve dissolution of one phase and nucleation and growth of another. While reactions in hydrothermal and metamorphic systems fall into this category, igneous processes dominantly involve nucleation and growth alone. In the metamorphic temperature range (~ 200–$750°C$) experiments have been performed in which equilibrium was investigated by determination of dissolution or growth of one of the minerals as a large single crystal in the reacting assemblage (Ch. 2). Wood & Walther (1983) showed that many of these data are consistent with Equation 9.9 and that the net rate is generally a linear function of temperature on both sides of the equilibrium boundary, i.e., in both dissolution and growth regimes. By assuming that the slow step in these experiments corresponds to reaction at the surface of the single crystal, Wood & Walther extracted rate constants using Equation 9.9. When data for a wide range of different mineral species are plotted on an Arrhenius

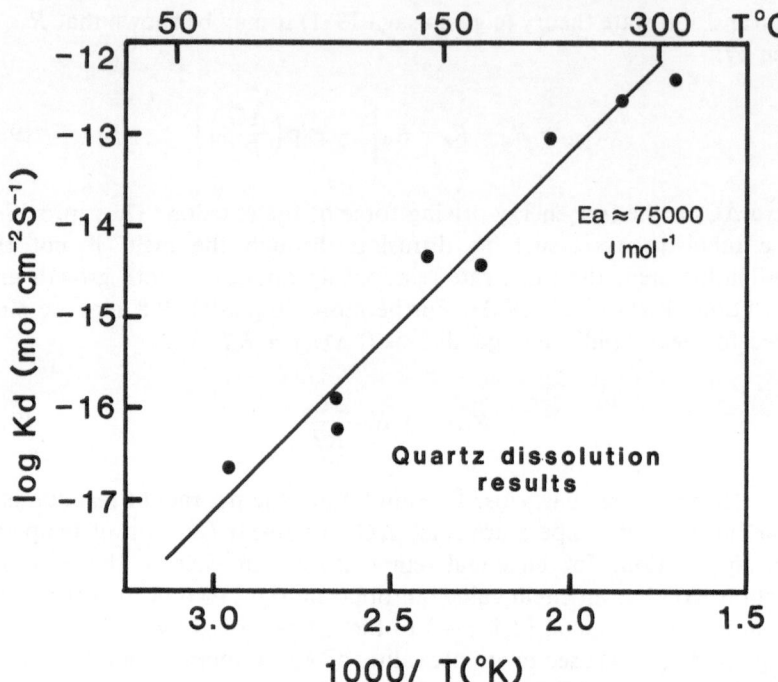

Figure 9.12 Arrhenius plot of dissolution rate constants for quartz using data of Rimstidt and Barnes (1980).

diagram (Fig. 9.13), it can be seen that many minerals dissolve and grow at similar rates. The rate constants are also consistent with the low temperature dissolution data of Rimstidt & Barnes (Fig. 9.12). It appears, therefore, that we have a reasonable semi-quantitative model for the rates of attachment and detachment at mineral/fluid interfaces under metamorphic conditions. These might be used, for example, to calculate the approximate time taken for a reaction to go to completion (complete dissolution of reactants and growth of products) for a given amount of free energy (or temperature) overstep (ΔG_r) and average grain size.

9.5 Elastic properties

The major part of this book deals with the application of experimental methods to determining the origin and evolution of rocks exposed at the Earth's surface. Such rocks only represent rather scattered samples of the upper 150–200 km of the Earth, however, and there is no way that one can obtain samples from deeper down. Despite this, the physical properties of the mantle and core may be studied in detail by determining the way in

162

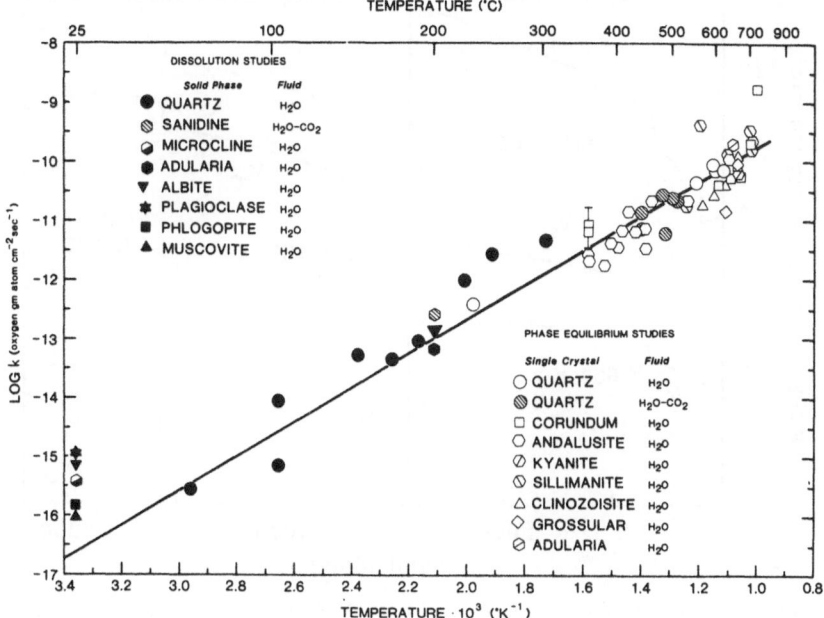

Figure 9.13 Arrhenius plot of the reciprocal of absolute temperature $(1/T)$ versus the logarithm (to the base 10) of the rate constant, determined from both dissolution and phase-equilibrium studies. From Wood and Walther (1983).

which they propagate seismic waves. One can then attempt to constrain the chemical and physical make-up of these regions by comparing observed physical properties with laboratory measurements on those high pressure phases that are postulated to exist at great depth.

By 1937, seismologists had a good idea of the variation of compressional (P-wave) and shear (S-wave) velocity with depth in the Earth (Fig. 9.14). However, the application of these data to interpretation of the chemical and mineralogical constitution of the mantle and core requires other information. First, the Earth's mass and moment of inertia, which are well known, place important constraints on density and mass distribution. Second, it is necessary to have equations of state (volume as a function of P and T) for the silicate, oxide, and metal phases which are candidate materials for the deep Earth. Birch (1952) used finite strain theory to derive a general equation of state for solids at very high pressures. He showed that, to a good approximation, most materials could be described by a pressure–density equation of the form:

$$P = \frac{3}{2} K_T \left[\left(\frac{\rho}{\rho_0} \right)^{7/3} - \left(\frac{\rho}{\rho_0} \right)^{5/3} \right] \qquad (9.10)$$

163

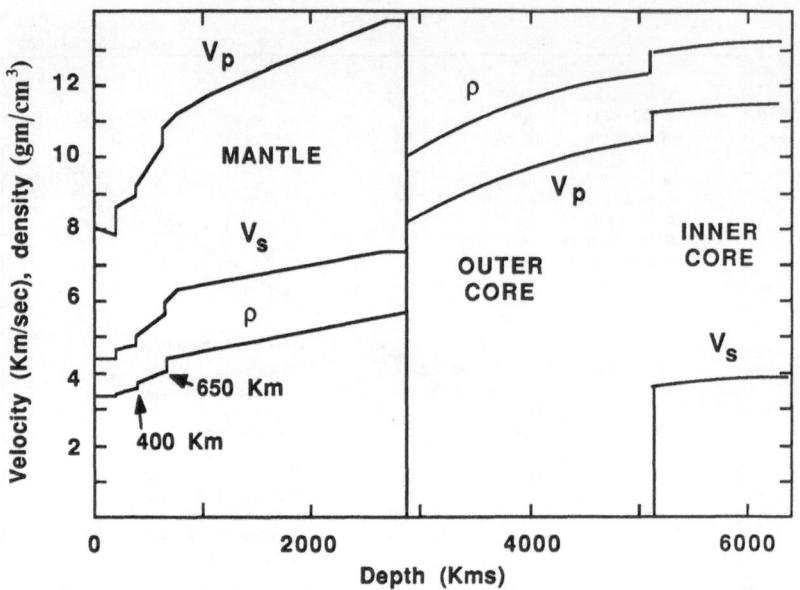

Figure 9.14 P- and S-wave velocities (V_p and V_s) and density (ρ) as a function of depth in the Earth.

where P is pressure, ρ is density at pressure P, ρ_0 is density at zero pressure and K_T is the zero pressure isothermal bulk modulus. A lot more can be done with Equation 9.10 if a few thermodynamic relationships are considered. The isothermal bulk modulus K_T is the reciprocal of the compressibility β of the material at temperature T.

$$K_T = \frac{1}{\beta} = -V\left(\frac{\partial P}{\partial V}\right)_T$$

Thus K_T can be derived by measuring the volume as a function of pressure in the way that Birch did. The isothermal bulk modulus is related to the adiabatic (constant entropy) bulk modulus K_S by:

$$K_S = K_T(1 + T\alpha\gamma_{th}) \tag{9.11}$$

where T is temperature, α is the coefficient of thermal expansion and γ_{th} is the thermal Gruneisen parameter. The temperature-dependent term on the right hand side of Equation 9.11 is only a few per cent of the total, so that K_S is not too different from K_T. The dimensionless Gruneisen parameter is defined as follows:

$$\gamma_{th} = \frac{\alpha K_S}{\rho C_p} \quad (\approx 1 \text{ for silicates})$$

164

Where C_p is the heat capacity at constant pressure. The adiabatic bulk modulus can also be derived from the P- and S-wave velocities V_P and V_S since by definition:

$$\frac{K_S}{\rho} = (V_P^2 - \tfrac{4}{3} V_S^2) = \phi \text{ (the seismic parameter)} \qquad (9.12)$$

Thus, seismic velocities, density, and bulk modulus are all tied together through Equations 9.10, 9.11, and 9.12. In principle then, one can obtain seismic velocities at low pressure and use Birch's equation of state to predict what they should be at high pressure. Alternatively, measurement of density as a function of pressure may be used to obtain bulk modulus and hence the seismic parameter. Birch used the equation of state to show that, with any reasonable set of assumptions, the Earth, and in particular the 400 to 1,000 km depth transition zone could not be a homogeneous solid under self-compression. This supported the idea that the seismic discontinuities at 400 and 650 km (Fig. 9.14) had to be due to changes either of phase or of chemical composition.

Since this early paper, there have been many studies aimed at improving the data base for high P–T extrapolation of density and seismic data. Direct measurement of sound-wave velocities is generally made with an experimental arrangement of the type shown in Figure 9.15. A transducer bonded to the sample oscillates at ultrasonic frequencies (about 20 MHz), and the velocities of such waves through the sample are determined by finding the conditions for interference between waves reflected from the sample's free and fixed ends. Experiments may be made at pressures up to 30 kbar and temperatures up

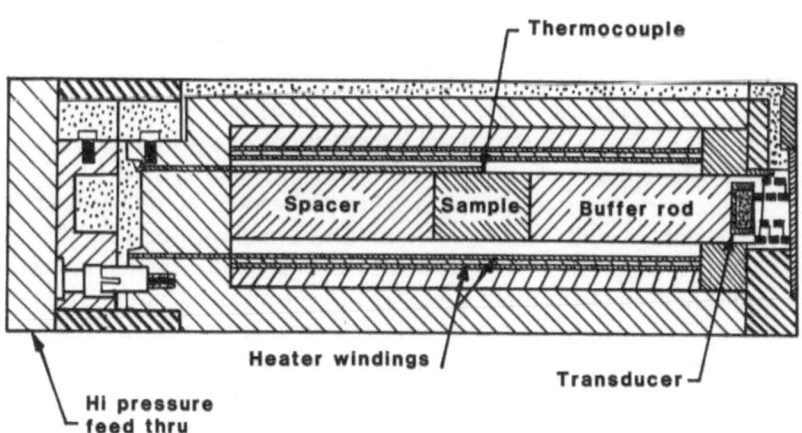

Figure 9.15 Ultrasonic interferometry. Cross-section through furnace with sample and buffer-rod in place. (Spetzler 1970).

to 1,000°C by placing the sample inside an internally heated or solid-media pressure vessel (Ch. 3). Results of P- and S-wave velocity measurements on polycrystalline aggregates yield bulk modulus and the modulus of rigidity (ρV_s^2). If the sample is a single crystal, individual elastic moduli are obtained by measuring velocities in different crystallographic directions. The bulk modulus is then computed from an appropriate average of the individual moduli.

Measurements of single crystal elastic properties are generally regarded as superior to those on polycrystalline aggregates. In the former case it is not necessary to make any correction for residual porosity of the sample. Furthermore, measurements of velocities in different crystallographic directions enables the possibility of velocity anisotropy within the Earth to be considered. Such anisotropy would arise, for example, if anisotropic crystals were oriented preferentially in some particular direction. Despite their utility, however, there are some difficulties with single crystal ultrasonic measurements, particularly when applied to high-pressure minerals. First, it is necessary to set up experiments for a number of different crystallographic orientations. (A minimum of three are required for a cubic crystal, and more are needed for minerals of lower symmetry). Second, sample crystals on the order of 5 mm on each side are required. While naturally occurring crystals of such dimensions may be found in rocks, it is extremely difficult to prepare synthetic ones that are big enough. So, single crystal ultrasonic studies can be made on near-surface materials but not on anything which might occur at great depth and which has to be synthesized at high P and T. Fortunately, another technique, that of Brillouin scattering (Weidner *et al.* 1975), has been developed to enable measurement of single crystal elastic properties on small crystals of about 100 μm in size. Crystals of this size can often be made at high pressure.

Brillouin scattering relies on the fact that sound waves passing through a crystal cause fluctuations in its refractive indices. Any light scattered by the crystal undergoes a frequency shift due to the frequency of the refractive index fluctuations. The experimentalist sets up a geometry in which monochromatic (laser) light is incident on the crystal. Observations of the frequency shifts in scattered light under conditions of constructive interference, much like Bragg's Law for constructive interference in X-ray diffraction, are then made. The frequency shift gives the velocity of the acoustic sound wave in the crystal. The method has a couple of advantages apart from the fact that small crystals can be used. For example, it is not necessary to generate an acoustic wave in the specimen. Such sound waves are produced continuously in any solid from normal atomic vibrations. In addition, it is not necessary to prepare different samples for the measurement of velocities in different crystallographic directions. The crystal is simply rotated in the light beam.

Sound velocity measurements provide an important basis for understanding the interior constitution of the Earth. Most of the measurements have been made at low pressures and temperatures, however, and there are problems in extrapolating them to, say, 200 kbar and 2,000°C. Although equations of state like that of Birch provide us with a model for extrapolation, it is still really important to know both the temperature and pressure derivatives of K_T. Birch's equation (9.10) has, built into it, the assumption:

$$K_T' = \left(\frac{\partial K_T}{\partial P}\right)_T = 4.0$$

which is actually pretty close to the measured value for many silicates (Jeanloz & Thompson 1983). Furthermore, it can easily be modified for other values of K_T' if $P-V-T$ data are good enough to obtain this parameter.

It appears that sound velocity measurements need to be complemented by precise $P-V-T$ data under deep mantle conditions in order to distinguish between a number of models for deep Earth structure. One way in which this might be done is by the accurate determination of crystal lattice parameters by *in situ* X-ray diffraction at high pressures and temperatures.

Combined X-ray and high $P-T$ studies have been performed with some success in the diamond anvil apparatus (Ch. 3). Better $P-T$ control is obtained with the cubic anvil device (Ch. 3) but, in order to get the X-rays to the sample, appropriate windows must be made. Fortunately, it is not necessary to scan a detector around the sample as in normal X-ray diffraction. A white X-ray source with a wide spectrum of wavelengths (e.g., synchrotron radiation) is used instead of the conventional single-wavelength source (e.g., Yagi & Akimoto 1982, 1984). The detector is set at a fixed diffraction angle and the wavelengths that fulfill the Bragg condition ($n\lambda = 2d \sin \theta$) are determined for the fixed value of theta. This yields an energy-dispersive diffraction pattern without the need to move the detector. The exposure time per sample can be cut to just a few minutes or even seconds if the extremely high intensity synchrotron radiation is used as the white X-ray source. Use of this source minimizes problems associated with keeping P and T stable during collection of the data. We envisage that application of these methods will yield a considerable quantity of precise high $P-T$ volume data in the near future. They will result in better values of density, K_T and the derivatives of K_T with respect to P and T than are available at present.

9.6 Conclusions

This chapter has been a quick trip through the measurement of physical and transport properties of minerals and melts. Much remains to be done. Melt densities and viscosities are only known well at atmospheric pressure, with a few compositions having been studied at pressures up to 20 kbar. Despite the fact that high-temperature creep is controlled by diffusion and vacancy concentrations, little work has been done on the dependence of deformation on oxygen fugacity. Direct determination of $P-V-T$ relationships at very high pressures is only just becoming possible. These and the other problems discussed in this chapter will be keeping experimentalists at work for a long time to come.

Appendix

A.1 Measuring temperature

Temperature can be controlled and measured with high precision, ± 1 to $5°C$, under most of the experimental conditions described in this book. This is done with the aid of three types of welded thermocouple, chromel–alumel, platinum–rhodium, and tungsten–rhenium.

A.1.1 Chromel–alumel

Spools of wire for chromel–alumel thermocouples are commercially available and relatively cheap. The two legs of the thermocouple are passed down two-hole thermocouple ceramic and welded or melted together under anoxic conditions (see under tungsten–rhenium below). Alternatively, ready-made sheathed thermocouples can be purchased. If ready-made, they may already be calibrated, otherwise the spools of wire should be calibrated before use. This is usually done by making a thermocouple and checking the melting point of sodium chloride ($T_m = 800.5°C$). Chromel–alumel melts at about $1,250°C$ and oxidizes fairly rapidly in air above $1,150°C$, so this thermocouple is restricted to use below about $1,150°C$ in either oxidizing or reducing atmosphere. Use at temperatures above $1,050°C$ requires frequent recalibration.

A.1.2 Platinum–rhodium

Various alloys of platinum and rhodium are used in thermocouples. In the United States the commonest is one leg of pure Pt and the other of $Pt_{90}Rh_{10}$. In Europe and Japan, $Pt_{87}Rh_{13}$ replaces the Rh_{10} leg. Higher temperature thermocouples are typically made from more rhodium-rich alloys such as $Pt_{70}Rh_{30}$ combined with $Pt_{94}Rh_{6}$. The thermocouple legs are easily arc-welded together in air. Calibration is typically made at the melting point of gold. Gold wire is suspended between two otherwise unconnected legs and the three-wire mixture heated until the gold melts at which point the thermocouple emf is lost.

Pt–Rh thermocouples work very well except in reducing carbonaceous environments such as those in the piston–cylinder apparatus. If used in the piston-cyclinder, the thermocouples poison rapidly at $1,200°C$ and above, so that the emf drifts downwards with time.

169

At high temperatures of 1,600°C and above, the $Pt_{100}/Pt_{90}Rh_{10}$ thermocouple starts to drift even in air due primarily to changes in the Seebeck emf of platinum. This drift may be reduced by using rhodium alloys on both legs, e.g., $Pt_{70}Rh_{30}/Pt_{94}Rh_6$ or $Pt_{60}Rh_{40}/Pt_{90}Rh_{10}$ produce thermocouples in which the emf is stable for many hours even at 1,700°C. However, frequent recalibration is required for accurate work above 1,500°C.

A.1.3 Tungsten–rhenium

Tungsten–rhenium thermocouples can only be used under reducing conditions, but they are extremely stable to high temperatures, up to 2,200°C under such conditions. The commonest alloys in use are $W_{97}Re_3$ as one leg and $W_{75}Re_{25}$ as the other or $W_{95}Re_5$ combined with $W_{74}Re_{26}$. The wires may be arc-welded together under anoxic conditions. This is achieved by flowing a stream of nitrogen or argon across the contact between the two wires as the arc is struck. Because of their stability under reducing conditions, tungsten–rhenium thermocouples have replaced platinum–rhodium in most piston-cylinder laboratories. The W–Re wires show more batch-to-batch variation than do other thermocouple types, and should be calibrated before use.

A.2 Measuring and generating pressure

A.2.1 Pressure media

The requirements for an ideal medium are:

(a) that it is hydrostatic, that is, it does not possess any shear strength;
(b) that it does not react with the pressure system, furnace, or sample capsule materials it contacts; and
(c) that it has a low compressibility.

In practice there is no ideal medium. The most commonly used media are water, argon gas, and nitrogen gas. Water is used because it has a low compressibility and is relatively unreactive in externally heated vessels. Its low compressibility allows for very simple pressure generating systems. Argon is unreactive but, because of its relatively high compressibility, requires more complex pumping systems. Argon freezes at about 13 kbar and room T, so nitrogen is used at greater pressures although it is considerably more compressible. None of the media used in solid-media vessels are completely without shear strength, and so cannot generate truly hydrostatic pressures.

A.2.2   Pressure measurement

The major difficulty with pressure measurement is accurate calibration. Primary pressure calibration is accomplished using a device called a dead weight guage which consists of a freely floating piston in a cylinder containing the pressure. Weights are placed on the piston to balance exactly the force generated by the pressure. The weight is used to calculate the force on the piston, and the pressure is determined by dividing that force by the area of the piston. The only trick to this technique is making a frictionless seal on the piston. This becomes progressively more difficult as pressure increases, but has been used successfully at pressures up to 25 kbar. At greater pressures the force per unit area method may still be used, but some means must exist to evaluate friction. This has been done in piston-cylinder, solid-media devices by measuring a reaction boundary using "piston-in" and "piston-out" experiments (Johannes *et al.* 1971). Pressures beyond the range of single-stage piston-cylinder vessels can only really be calibrated by the extrapolation of some physical property, for example, the change in unit cell edge of NaCl or gold.

A.2.3   Pressure generation

Except for shock experiments, all high pressure systems rely on the area ratio technique to multiply pressure. If two static pistons are in contact on their faces, then the force on one face is equal to that on the other face,

$$F_1 = F_2$$

and because $F = P \cdot A$,

$$P_1 \cdot A_1 = P_2 \cdot A_2$$

or

$$P_2 = P_1(A_1/A_2)$$

So if $A_1 > A_2$ then pressure will be multiplied by the ratio of the areas. This principle is used in intensifiers for fluids, and in piston-cylinder, diamond-anvil, and multi-anvil devices.

A.3   Controlling oxygen fugacity

There are three methods in use for controlling oxygen fugacity in experiments; gas mixing, solid buffers, and hydrogen membranes. The first is used only at atmospheric P and the second two are used only at higher P.

171

A.3.1 Gas mixing

This technique was developed by Darken & Gurrey (1945) and has been used extensively to study melting relations in iron-bearing systems. The technique relies on gas-phase reactions between two "feed" gases, usually CO/CO_2 or H_2/CO_2. The reaction for CO/CO_2 is:

$$CO + \tfrac{1}{2}O_2 = CO_2 \qquad \text{(A.1)}$$

in which f_{O_2} is fixed as follows:

$$f_{O_2} = \left(\frac{P_{CO_2}}{P_{CO} \cdot K_1}\right)^2 \qquad \text{(A.2)}$$

Because the reaction occurs at atmospheric P and high T, partial pressures may be substituted for fugacities. The values of K_1 can be found in tables of standard thermodynamic properties. In practice the partial pressures are controlled by adjusting the flow rate of the two feed gases.

A.3.2 Solid buffers

The use of "solid" oxygen buffers in high pressure experiments was invented by Eugster (1957) and has been used extensively in cold-seal, gas-vessel, and piston-cylinder experiments. The technique is based on the use of a crystalline phase assemblage to fix f_{O_2} and the existence of a capsule material which acts as an ideal semipermeable membrane to H_2. Some of the crystalline phase assemblages used are listed in order of decreasing f_{O_2} below:

$$2Fe_3O_4 + \tfrac{1}{2}O_2 \rightleftharpoons 3Fe_2O_3$$

$$Ni + \tfrac{1}{2}O_2 \rightleftharpoons NiO$$

$$3Fe_2SiO_4 + O_2 \rightleftharpoons 2Fe_3O_4 + 3SiO_2$$

$$Fe + \tfrac{1}{2}O_2 \rightleftharpoons FeO$$

Each of the above reactions will fix f_{O_2} in a system at constant P and T as long as all crystalline phases in the reaction are present. For instance, if crystalline Ni and NiO are present then f_{O_2} is given by:

$$f_{O_2} = (1/K)^2$$

where K is the equilibrium constant for the reaction as written. In practice, these reactions occur in the presence of H_2O, and it is H_2 which actually controls the system since this is the species which may diffuse from buffer to sample. The above reactions should therefore be rewritten as:

$$2Fe_3O_4 + H_2O \rightleftharpoons 3Fe_2O_3 + H_2$$

$$Ni + H_2O \rightleftharpoons NiO + H_2$$

$$3Fe_2SiO_4 + 2H_2O \rightleftharpoons 2Fe_3O_4 + 3SiO_2 + 2H_2$$

$$Fe + H_2O \rightleftharpoons FeO + H_2$$

Each of the above reactions results in fixed H_2 fugacity provided that H_2O activity is fixed. H_2O activity will be fixed if the fluid phase contains only H and O. Under the conditions of these buffers O_2 concentrations are trivial so H_2 and H_2O are the only important species and total $P = P_{H_2} + P_{H_2O}$. The activity of H_2O will also be fixed in more complex fluids provided there is one crystalline phase present for each additional fluid component added. For instance C–O–H fluid in the presence of graphite, or C–O–H–S fluids in the presence of graphite and pyrrhotite ($Fe_{1-x}S$). The solid buffer assembly and desired fluid are placed in an outer capsule, and the sample and desired fluid are placed in an inner capsule. The commonly used capsule materials, platinum or alloys of palladium–silver, are very permeable to H_2 but impermeable to all other gases. Thus if one of these metals is used for the inner capsule, the H_2 activity generated in the buffer assemblage is perfectly transmitted to the inner capsule. That H_2-together with H_2O in the inner capsule, react to fix f_{O_2} through the H_2O-forming reaction:

$$H_2 + \tfrac{1}{2}O_2 \rightleftharpoons H_2O \qquad\qquad (A.3)$$

which fixes f_{O_2} as follows:

$$f_{O_2} = (f_{H_2O}/f_{H_2} \cdot K)^2 \qquad\qquad (A.4)$$

To recapitulate, the buffer assemblage in the outer capsule fixes H_2 activity which is transmitted through the wall of the inner capsule where it reacts with H_2O to fix f_{O_2}. These techniques are discussed in more detail by Eugster & Skippen (1967) and Huebner (1971).

The solid buffer systems have two limitations: they are consumed rapidly at high temperatures, and the O_2 fugacities they produce are spaced at wide intervals. These limitations are each overcome with the H_2 membrane described next.

A.3.3 The Shaw hydrogen membrane

Shaw (1963a, 1967) extended the concept demonstrated by Eugster (1957) by replacing the buffer assemblage in the outer capsule with a palladium–

silver tube attached to an H_2 reservoir outside the pressure vessel. The fugacity of H_2 inside the vessel could then be controlled by adjusting the H_2 pressure in the external system. Because the fugacity coefficient of pure H_2 is well known (and nearly unity) at the low P, high T conditions encountered (for instance, the H_2 pressure at 2 kbar total P, 800°C, and an f_{O_2} at NiNiO is only about 6 bar), the fugacity of H_2 in the vessel may be accurately calculated. Just as in the case of solid buffers, the f_{O_2} inside the sample capsule is fixed by reaction between H_2 and H_2O as given by Equation A.4. The efficiency of the membrane and solid buffering techniques is controlled by the permeability of the membrane and inner capsule material to hydrogen. The order of permeabilities is $PdAg > Pt \gg Au > Ag$, and permeability varies with T and, in the case of alloys, with alloy composition. For more information the reader is directed to Gunter *et al.* (1979, 1987).

The main limitations of the membrane technique are first, that no one has managed to put a membrane inside a solid-media vessel and second, that to generate low O_2 fugacities requires very high hydrogen pressures. The great advantages of membranes include the ability to vary H_2, and hence O_2 fugacity continuously, and the fact that there is no buffer assembly to be exhausted in long runs.

A.4 Materials

In this section brief descriptions of some of the most commonly used materials are given together with some criteria for selecting them. The account is not meant to be exhaustive, rather to provide a feeling for the properties.

A.4.1 Furnaces

The furnaces used in vessels are all of the electrical resistance type and consist of two parts, the electrical conductor (element) and the insulation. The important characteristics of the element are: upper T limit, type of atmosphere it can withstand (reducing, oxidizing), electrical resistance, and the change in resistance with T. The electrical resistance is important because low resistances result in high electrical currents that may be too large for electrical leads going inside pressure vessels. The change in resistance with T is important because elements with large changes commonly produce large T gradients, while elements which show little change in resistance with T can be used to produce large isothermal zones. In general, pure metals have low resistances at low T, and resistance increases rapidly with increasing T. The converse is true of many complex alloys such as the FeNiCrAl type with the trade name Kanthal A1.

The important characteristics of insulators include the upper T limit, the electrical and thermal resistance, the ability of the material to be shaped (by casting or machining), and the effects of the pressure medium on the material (for instance, some insulators dissolve argon gas under high $P–T$ conditions and disintegrate when the gas forcefully exsolves on quenching).

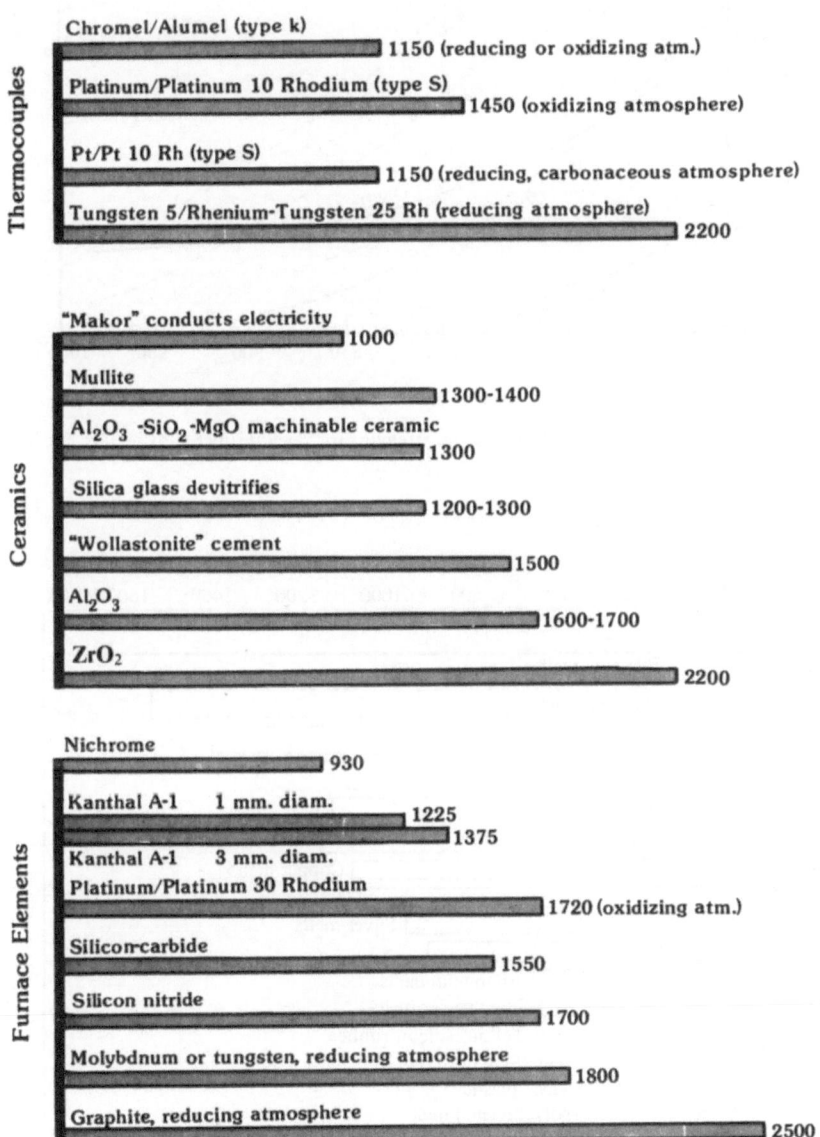

Figure A.1 Temperature ranges for some thermocouple, ceramic, and furnace element materials.

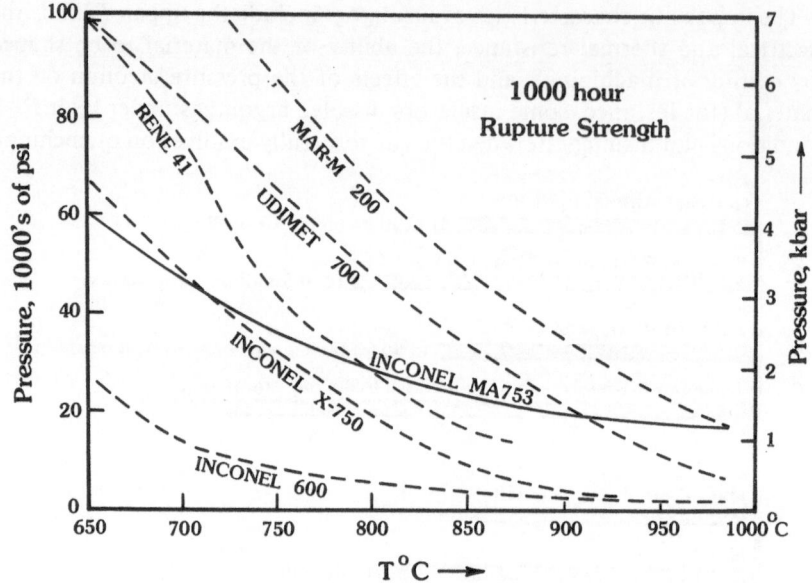

Figure A.2 Comparative high temperature strengths for several super alloy metals. See Table A.1 for alloy compositions.

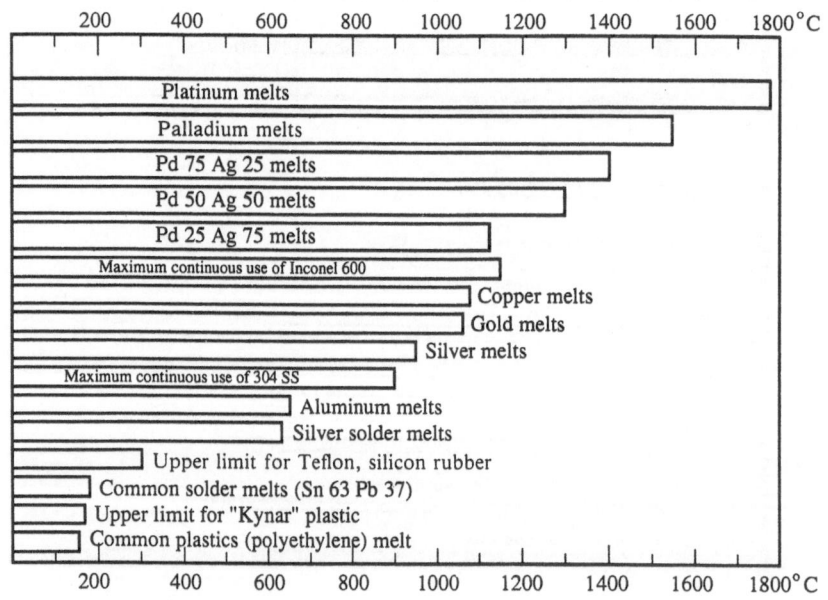

Figure A.3 Melting points and working limits of capsule and other materials.

Some characteristics of element and insulation materials are given in Figure A.1.

A.4.2 Other materials

The only type of vessel material discussed here is that for use in externally heated vessels. Other applications are too specialized. The most important aspect of material for external heating is its strength at high T. By far the most commonly used material has been Rene 41, but as can be seen in Figure A.2, there are other alloys available with higher strengths.

The commonly used metals for sample capsules are shown in Figure A.3. The important characteristics of capsule materials in addition to their upper T limit are hydrogen permeability and iron solubility.

The thermal stabilities of several other common materials are shown in Figure A.3

Table A.1 Metal compositions (wt%).

Alloy	Mn	Fe	Cr	Ni	Co	Mo	W	Ti	Al
Inconel 600	0.5	8	15.5	76	–	–	–	0.35	0.25
Inconel X750	0.5	7	15.5	73	1	–	–	2.5	0.7
Rene 41	–	–	19	55	11	10	–	3.1	1.5
Udimet 700	–	–	15	53	18.5	5.2	–	3.5	4.3
MAR-M 200	–	–	9	60	10	–	12	2	5
Inconel MA753†	–	–	20	75	–	–	–	2.5	1.5

†Includes 1.3% Y_2O_3

A.5 Preparation of starting materials

The choice of starting materials can have dramatic effects on the results of an experiment. Therefore, it is important to emphasize again, following Fyfe (1960), that equilibrium can only be demonstrated by proving that product and reactant assemblages have the same free energies under a given set of $P-T$ conditions. There are several approaches to such proofs, reversal being one, solubility in another phase (e.g., cryolite melt; Weill 1966) another. In either case, however, it is often necessary to synthesize the phases of interest before performing the experiment. Let us assume therefore that we intend to make some synthetic minerals and melts for the purpose of performing a phase equilibrium study.

Starting materials for synthesis should be as reactive as possible and as chemically homogeneous as possible to ensure that a homogeneous product is readily produced. Gels, glasses, and freeze-dried oxide mixes meet these criteria.

A.5.1 Gel preparation

Coprecipitated gels are made by putting all the elements of interest into acid solution then coprecipitating them with an SiO_2 gel by making the solution alkaline (Hamilton & Henderson 1967). High purity or analytical grade chemicals are used. Sodium, potassium, and calcium carbonates (Na_2CO_3, K_2CO_3, $CaCO_3$) are dried at $110°C$, weighed and transferred to a 500 ml Teflon beaker. Aluminum, iron, and magnesium are weighed and transferred to the beaker as metals. Titanium and chromium may be added as tetraethyl orthotitanate and ammonium dichromate solutions, respectively, just before the coprecipitation stage.

When all of the powders are in the beaker, it is covered with a watch glass (to prevent loss through effervescence) and 50% nitric acid added *slowly* to dissolve them. Aluminum and iron dissolve very slowly, so such compositions have to be immersed (still covered) in a water bath for a few hours. The solution is then evaporated nearly to dryness to remove excess nitric acid and the residue redissolved in a small volume of water.

The beaker is now placed in an ice bath to keep it cold. When the solution is cold, SiO_2 is weighed into a glass tube as liquid tetraethyl orthosilicate (TEOS). The liquid is then washed into the beaker with ethyl alcohol, stirred (Teflon stirrer) and alcohol added until the solution is completely miscible. Concentrated ammonium hydroxide (density 0.88) is now added to precipitate the silica. The solution is stirred continuously and kept cool in the ice bath. Since the neutralization generates heat, it is essential that the solution be kept cool, otherwise the TEOS may evaporate. The gelling process continues until the stiff, viscous gel smells strongly of ammonia. The gel is allowed to stand overnight at room temperature (to ensure complete precipitation) before drying on a water bath and then in an oven at $110°C$. The dried gel is ground in an agate mortar and transferred to a covered platinum crucible. The nitrates are decomposed very slowly over a Bunsen burner, and the gel is fired in a furnace at about $800°C$ for a few hours. It is then cooled and stored in a dessicator.

One or two points need to be made. A small amount (1–2%) of evaporation of TEOS often occurs during the process. Most workers test the SiO_2 yield with their procedures and add the appropriate amount of additional TEOS. After firing, the gel should be completely amorphous and very reactive. Any Fe or Cr will will be present as oxidized Fe(III) or Cr(VI) respectively. The gel can be reduced to other oxidation states by running it at $800°C$ under appropriate conditions in a controlled f_{O_2} furnace.

Another gel preparation technique is described by Mukherjee (1984) in which all metals (Si, Na, Al, etc.) are added as metallo-organic complexes. The solutions are gelled by slow heating and evaporation and can be made directly into hydrous "glasses." These glasses have many advantages for hydrous experiments.

A.5.2 Glass preparation

Glasses are generally less reactive than gels, but many people find it easier to make them with the correct composition. In addition, phase equilibrium studies involving liquids are better made with a supercooled liquid (i.e., glass) starting material than with a gel. If gels are already available, and they are iron-free, they may be turned into glasses by melting in a platinum crucible for about 15 min and then cooling as rapidly as possible (e.g., on an ice-pack). The required temperature depends on composition and can be determined from the phase diagram, but $1,400-1,500°C$ was good enough for most we have made.

Glasses may be made directly from oxides and carbonates. Dry (at $110°C$) Na_2CO_3, K_2CO_3, and $CaCO_3$ are generally used for Na, K, and Ca. SiO_2, Al_2O_3, MgO, Cr_2O_3, TiO_2, Fe_2O_3 are added as high-purity oxides which have been heated ($\sim 1,000°C$) for a few hours to remove all moisture. The oxides and carbonates are weighed and transferred to an agate mortar, ground under acetone, and homogenized as thoroughly as possible. When dry, the mixture (if iron-free) is transferred to a covered platinum crucible and heated, first at $800°C$ for a few hours to remove most of the CO_2, then above the liquidus for about 15 min. The melt is then quenched, broken out of the crucible, and reground. Glasses made directly from oxides are usually remelted at least once or twice (and eight times may be necessary) after the initial treatment to ensure homogeneity (Schairer 1959).

The only major problems which can occur with glass making are alkali loss and Fe(0)–Fe(II)–Fe(III) equilibrium. Alkali loss will not generally be significant with two 15 min fusions in air at temperatures a little above the liquidus. Some compositions may require three or four longer fusions to get homogeneity, however, and in these cases alkali loss can occur. In addition, if melting is done at low f_{O_2}, Na and K loss are enhanced. Fortunately, glasses are easy to analyze, either by microprobe or by wet-chemical methods, so that compositions and homogeneity may be checked. Iron-bearing glasses present more of a problem. If Fe_2O_3 is used as the source of iron, and the mixture is melted in air, then magnetite or an aluminous spinel often precipitates, and it can be very difficult to remove. In general, however, one is trying to make glasses in which most or all of the iron is present as Fe(II). This may be done in a gas-mixing furnace by running above the liquidus at an f_{O_2} in the wustite stability field. Under these conditions, iron dissolves readily into platinum so an alternative crucible material must be used. The best alternative is probably molybdenum crucibles which are easy to machine from Mo rod, and which do not interact strongly with Fe-bearing melts at low f_{O_2}. If a controlled atmosphere furnace is not available, then it is still possible, but a little trickier, to obtain Fe(II)-bearing glasses. In this case, a thick-walled graphite crucible with a tightly fitting graphite lid is made. The Fe_2O_3-bearing mixture is loaded into the crucible and the latter placed in a furnace at 1,350–1,400°C and 1 atm. Despite the absence of a low f_{O_2}-atmosphere, the ferric iron reacts with the carbon to yield ferrous iron in a greeny-brown glass. Depending on dimensions of crucible and sample, complete conversion takes 10 min to 1 hour. It is necessary to calibrate the time required carefully, for a few minutes too long in the furnace will produce iron metal in the glass.

A.5.3 Freeze-dried mixes

Freeze-drying, a variant of the gel technique, is described by Finnerty et al. (1978). All of the elements are put into aqueous solution by dissolving a carboxylic acid salt in water, e.g., magnesium citrate for Mg, calcium formate for Ca, and aluminum formyl acetate for Al. Silica is added as an aqueous sol such as Ludox AS (DuPont de Nemors). Since Ludox contains a small amount of sodium, Finnerty et al. suggest ion-exchanging a 1 M SiO_2 solution with NH_4^+ on an exchange column before use. The aqueous solution is frozen rapidly by squirting through a fine capillary tube into continuously stirred liquid nitrogen. The frozen droplets are then placed in a laboratory freeze-drier, and water is removed by vacuum sublimation. The final step is decomposition of the organic species in air at about 600°C to yield an intimate homogeneous mix of oxides. Although not yet tried by the authors, this method seems to yield products of comparable reactivity and homogeneity to those obtained by gelling.

A.5.4 Oxide mixes

Mixtures of oxides are much less reactive than the amorphous materials discussed in the previous sections. They may, however, be useful for synthesis at high temperatures, for synthesis in the presence of abundant H_2O fluid, or both. One might use them in preference to glass or gel for synthesis of ferrous silicates, since amorphous starting materials containing Fe(II) can be difficult to make. For high temperature synthesis in a controlled f_{O_2} furnace a mixture of oxides, Fe metal and Fe_2O_3 would be ground together under acetone, pressed into a pellet, and reacted at high-temperature. The pellet would then be reground and re-reacted two or three more times to improve homogeneity of the product. If a controlled atmosphere furnace were not available, then the pellet could be reacted in an evacuated sealed silica-glass tube.

Hydrothermal synthesis often works very well with oxide mixes. In this case, Fe and Fe_2O_3 are mixed in the sample so that the correct amount of total iron is present, but Fe metal is slightly in excess of the stoichiometric proportion. Sufficient water is added to the capsule so that the water will oxidize excess Fe to FeO, with the hydrogen passing out through the capsule walls. The sample is then run in a sealed gold or silver–palladium capsule under the desired P–T conditions.

A.5.5   Natural materials

Natural well characterized, well ordered minerals are generally favored as starting materials over poorly crystalline, disordered synthetics, particularly if they can be obtained as virtually pure end-members. They often can, and one can cite quartz, low-albite, jadeite, sillimanite, kyanite, adularia, anorthite, grossular, and muscovite as just a few examples of natural minerals used in well constrained experimental studies.

Finely ground rock powders are often used in melting and crystallization experiments. In the former case, they are brought directly to a P and T above the solidus and run, while in the latter they are taken above the liquidus, completely melted, and then brought back down temperature to the desired conditions.

A.5.6   Grinding and milling

All of the starting materials discussed here require extensive grinding or milling in order to homogenize them and to make them fine-grained and reactive. This cannot be done without putting a little of the mortar or mill material into the sample, so one needs to consider the effects such small compositional changes might have. Tungsten carbide or agate are generally good materials. The former does not generally add much contamination, and any that is present is very refractory and has little effect on silicate equilibria. Agate can contaminate the sample with SiO_2, but often a small excess of quartz is acceptable. If it is not, the sample can be made 1–2% deficient in SiO_2 during preparation. Steel should be avoided unless iron contamination is acceptable, in which case most of the steel can be removed with a magnet. Alumina and mullite add a small amount of Al_2O_3.

Bibliography/References

Ahrens, T. J. 1980. Dynamic compression of earth materials. *Science* **207**, 1035–41.

Aines, R. D., L. A. Silver, G. R. Rossman, E. M. Stolper, & J. R. Holloway 1983. Direct observation of water speciation in rhyolite at temperatures up to 850°C. *Abstract Geological Society of America* **15**, 512.

Akaogi, M., N. L. Ross, P. F. McMillan & A. Navrotsky 1984. The Mg_2SiO_4 polymorphs (olivine, modified spinel and spinel)- Thermodynamic properties from oxide melt solution calorimetry, phase relations, and models of lattice vibrations. *American Mineralogist* **69**, 499–512.

Akimoto, S. & M. H. Manghani (eds.) 1982. *High-pressure research in geophysics.* Tokyo: Center for Academic Publication.

Anderson, G. M. & C. W. Burnham 1965. The solubility of quartz in supercritical water. *American Journal of Science* **263**, 494–511.

Arculus, R. J. & J. W. Delano 1981. Intrinsic oxygen fugacity measurements: Techniques and results for spinels from upper mantle peridotites and megacryst assemblages. *Geochimica Cosmochimica Acta* **45**, 899–913.

Barnes, H. L. 1971. Investigations in hydrothermal sulfide solutions. In *Research techniques for high pressure and high temperature*, Ulmer, G. C. (ed.), 317–55. New York: Springer.

Barnes, H. L. 1963. Ore solution chemistry. I. Experimental determination of mineral solubilities. *Economic Geology* **58**, 1054–60.

Barrow, G. 1893. On an intrusion of muscovite-biotite gneiss in the South-East Highlands of Scotland. *Geological Society of London Quarterly Journal* **49**, 330–58.

Barrow, G. 1912. On the geology of lower Dee-side and the Southern Highland Border. *Proceedings of the Geologists Association* **23**, 268–84.

Barth, T. F. W. 1952. *Theoretical petrology.* New York: Wiley.

Bassett, W. A. 1979. The diamond cell and the nature of the earth's mantle. *Annual Review of Earth and Planetary Sciences* **7**, 357–84.

Berman, R. G., T. H. Brown, & H. J. Greenwood 1958. *An internally consistent data base for minerals – the system $Na_2O-K_2O-CaO-MgO-FeO-Fe_2O_3-Al_2O_3-SiO_2-TiO_2 -H_2O-CO_2$.* Atomic Energy of Canada Ltd Technical Rept TR 377.

Biggar, G. M. 1974. Oxygen partial pressures; control, variation, and measurement in quench furnaces at one atmosphere total pressure. *Mineralogical Magazine* **39**, 580–86.

Biggar, G. M. 1981. Comparison of $CO:CO_2$ with $H_2:H_2O$ gas mixtures for controlling gas fugacities in furnaces at one atmosphere. In *Progress in experimental petrology*, **5**, 88–9. National Environmental Research Council, Series D, No. 11.

Birch, F. 1952. Elasticity and constitution of the earth's interior. *Journal of Geophysical Research* **57**, 227–86.

Bischoff, J. L. & F. W. Dickson 1975. Seawater–basalt interaction at 200°C and 500 bars: Implications for origin of seafloor heavy metal deposits and regulation of seawater chemistry. *Earth and Planetary Science Letters* **25**, 385–97.

Bischoff, J. L. & W. E. Seyfried 1978. Hydrothermal chemistry of seawater from 25° to 350°C. *American Journal of Science* **297**, 838–60.

Boettcher, A. L. 1970. The system $CaO-Al_2O_3-SiO_2-H_2O$ at high pressures and temperatures. *Journal of Petrology* **11**, 337–79.

Boettcher, A. L & D. M. Kerrick 1971. Temperature calibration of cold-seal pressure vessels. In *Research techniques for high pressure and high temperature*, G. C. Ulmer (ed.), 179–93. New York: Springer.

Boettcher, A. L., K. E. Windom, S. R. Bohlen, & R. W. Luth 1981. Low-friction, anhydrous, low- to high-temperature furnace sample assembly for piston-cylinder apparatus. *Reviews of Scientific Instruments* **52**, 1903–4.

Bohlen, S. R. 1984. Equilibria for precise pressure calibration and a frictionless furnace assembly for the piston-cylinder apparatus. *Neus. Jahrbuch für Mineralogie* **1** Monetscheffe, 404–12.

Bohlen, S. R. & A. L. Boettcher 1981. Experimental investigations and geological applications of orthopyroxene geobarometry. *American Mineralogist* **66**, 951–64.

Bohlen, S. R. & A. L. Boettcher 1982. The quartz-coesite transformation: a precise determination and the effects of other components. *Journal of Geophysical Research* **87**, 7073–8.

Bohlen, S. R., E. J. Essene, & A. L. Boettcher 1980. Reinvestigation and application of olivine-quartz-orthopyroxene barometry. *Earth and Planetary Science Letters* **47**, 1–10.

Bourcier, W. L. & H. L. Barnes (1987). Rocking autoclaves for hydrothermal experiments I. Fixed volume systems. In *Hydrothermal experimental techniques*, G. C. Ulmer & H. L. Barnes (eds.), 189–215. New York: Wiley.

Bourcier, W. L., G. C. Ulmer, & H. L. Barnes (1987). Hydrothermal pH sensors of ZrO_2, Pd hydrides, and Ir oxides. In *Hydrothermal experimental techniques*, G. C. Ulmer & H. L. Barnes Jr. (eds.), 157–88. New York: Wiley.

Bowen, N. L. 1915. Crystallization of haplobasaltic, haplodioritic, and related magmas. *American Journal of Science* Series 4 **38**, 207–64.

Boyd, F. R. & J. L. England 1960. Apparatus for phase-equilibrium measurements at pressures up to 50 kilobars and temperatures up to $1,750°C$. *Journal of Geophysical Research* **65**, 741–8.

Buback, M. 1981. Spectroscopic investigations of fluids. In Nobel symposium on the chemistry and geochemistry of solutions at high temperatures and pressures. *Physics and Chemistry of the Earth* **13**, 345–60. New York: Pergamon Press.

Buback, M., D. A. Crerar, & L. M. V. Koplitz 1987. Vibrational spectroscopy of hydrothermal systems. In Ulmer, G. C. & H. L. Barnes Jr., *Hydrothermal experimental techniques*. 333–59. New York: Wiley.

Buening, D. U. & P. R. Buseck 1973. Fe–Mg lattice diffusion in olivine. *Journal of Geophysical Research* **78**, 6852–62.

Burnham, C. W. 1975. Water and magmas: A mixing model. *Geochimica Cosmochimica Acta* **39**, 1077–84.

Burnham, C. W. 1979a. The importance of volatile constituents. In *The evolution of the igneous rocks*, Yoder, H. S. (ed.), Princeton, N.J.: Princeton University Press.

Burnham, C. W. 1979b. Magmas and hydrothermal fluids. In *Geochemistry of hydrothermal ore deposits*, 2nd edn., H. L. Barnes, (ed.), 71–136. New York: Wiley.

Burnham, C. W. 1981. The nature of multicomponent aluminosilicate melts. In *Chemistry and geochemistry of solutions at high temperatures and pressures; physics and chemistry of the Earth 13*, D. T. Rickard & F. E. Wickman (eds.), 197–229. New York: Pergamon Press.

Burnham, C. W. and N. F. Davis 1971. The role of H_2O in silicate melts: I. $P-V-T$ relations in the system $NaAlSi_3O_8-H_2O$ to 10 kilobars and $1000°C$. *American Journal of Science* **270**, 54–79.

Burnham, C. W. & N. F. Davis 1974. The role of H_2O in silicate melts: II. Thermodynamic and phase relations in the system $NaAlSi_3O_8-H_2O$ to 10 kilobars, $700°C-1100°C$. *American Journal of Science* **274**, 902–40.

Burnham, C. W. & R. H. Jahns 1962. A method for determining the solubility of water in silicate melts. *American Journal of Science* **260**, 721–45.

Burnham, C. W. & H. Nękvasil 1986. Equilibrium properties of granite pegmatite magmas. *American Mineralogist* **71**, 239–63.

Burnham, C. W., J. R. Holloway, & N. F. Davis 1969a. *Thermodynamic properties of water to 1000°C and 10000 bars*. Geological Society of America Special Paper, **132**.

BIBLIOGRAPHY

Burnham, C. W., J. R. Holloway, & N. F. Davis 1969b. The specific volume of water in the range 1000–8900 bars, 20 to 900°C. *American Journal of Science* **267**, 70–95.

Calvet, E. & H. Prat 1954. *Microcalorimetre*. Paris: Masson.

Charlu, T. V., R. C. Newton, & O. J. Kleppa 1975. Enthalpies of formation at 970°K of compounds in the system $MgO-Al_2O_3-SiO_2$ from high temperature solution calorimetry. *Geochimica Cosmochimica Acta* **39**, 1487–97.

Chatterjee, N. D. & W. Johannes 1974. Thermal stability and standard thermodynamic properties of synthetic $2M_1$ muscovite, $KAl_2Si_3AlO_{10}(OH)_2$. *Contributions to Mineralogy and Petrology* **48**, 89–114.

Chou, I.-M. 1987. Oxygen buffer and hydrogen sensor techniques at elevated pressures and temperatures. In *Hydrothermal experimental techniques*, G. C. Ulmer & H. L. Barnes Jr. (eds.), 61–99. New York: Wiley.

Chou, I.-M. & J. D. Frantz 1977. Recalibration of Ag + AgCl buffer at elevated pressures and temperatures. *American Journal of Science* **277**, 1067–72.

Clemens, J. D. & V. J. Wall 1981. Origin and crystallization of some peraluminous granitic magmas. *Canadian Mineralogist* **19**, 111–31.

Clemens, J. D. & V. J. Wall 1984. Origin and evolution of a peraluminous silicic ignimbrite suite: the Violet Town volcanics. *Contributions to Mineralogy and Petrology* **88**, 354–71.

Crank, J. 1975. *The mathematics of diffusion*. Oxford: Clarendon Press.

Crawford, M. L. & L. S. Hollister 1986. Metamorphic fluids: the evidence from fluid inclusions. In *Fluid-rock interactions during metamorphism*, J. V. Walther & B. J. Wood (eds.). *Advances in Geochemistry* series, **5**, 1–35. New York: Springer.

Danielson, M. J., O. H. Koski, & J. Myers 1985. Recent developments with high temperature stabilized-zirconia pH sensors. *Journal of the Electrochemical Society* **132**, 296–301.

Darken, L. S. & R. W. Gurry 1945. The system iron-oxygen. I. The wustite field and related equilibria. *Journal of the American Chemical Society* **68**, 798–816.

Dickson, F. W., C. W. Blount, & G. Tunell 1963. Use of hydrothermal solution equipment to determine the solubility of anhydrite in water from 100°C to 275°C and from 1 bar to 1000 bars pressure. *American Journal of Science* **261**, 61–78.

Dirkse, T. P. & B. Wiers 1959. Stability and solubility of silver (II) oxide in alkaline solutions. *Journal of the Electrochemical Society* **106**, 284–87.

Donaldson, C. H. 1979. Composition changes in a basalt melt contained in a wire loop of $Pt_{80}Rh_{20}$: Effects of temperature, time, and oxygen fugacity. *Mineralogical Magazine* **43**, 115–19.

Douglas, T. B. & E. G. King 1968. High temperature drop calorimetry. In *Experimental thermodynamics, Vol. I: Calorimetry of non-reacting systems*, J. P. McCullogh & D. W. Scott (eds.), 293–332. London: Butterworth.

Drake, M. J. & J. R. Holloway 1981. Partitioning of Ni between olivine and silicate liquid: the 'Henry's law problem' reexamined. *Geochimica Cosmochimica Acta* **45**, 431–37.

Eggler, D. H. 1975a. CO_2 as a volatile component of the mantle: The system $Mg_2SiO_4-SiO_2-H_2O-CO_2$. *Physics and Chemistry of the Earth* **9**, 869–81.

Eggler, D. H. 1975b. Peridotite-carbonate relations in the system $CaO-MgO-SiO_2-CO_2$. *Carnegie Institute of Washington Yearbook* **74**, 468–75.

Eggler, D. H. 1978. The effect of CO_2 upon partial melting of peridotite in the system $Na_2O-CaO-Al_2O_3-MgO-SiO_2-CO_2$ to 35 Kb in a peridotite–H_2O-CO_2 system. *American Journal of Science* **278**, 305–43.

Eggler, D. H., B. O. Mysen, T. C. Hoering, & J. R. Holloway 1979. The solubility of carbon monoxide in silicate melts at high pressures and its effect on silicate phase relations. *Earth and Planetary Science Letters* **43**, 321–30.

184

Elliot, W. C., D. E. Grandstaff, G. C. Ulmer, T. Buntin, and D. P. Gold 1982. An intrinsic oxygen fugacity study of platinum–carbon associations in layered intrusions. *Economic Geology* **77**, 1493–1510.

Eugster, H. P. 1957. Heterogeneous reactions involving oxidation and reduction at high pressures and temperatures. *Journal of Chemical Physics* **26**, 1760–1.

Eugster, H. & G. B. Skippen 1967. Igneous and metamorphic equilibria involving gas equilibria. *Researches in Geochemistry* **2**, 492–526.

Ferry, J. M. & F. S. Spear 1978. Experimental calibration of the partitioning of Fe and Mg between biotite and garnet. *Contributions to Mineralogy and Petrology* **66**, 113–7.

Fine, G. & E. Stolper 1985. The speciation of carbon dioxide in sodium aluminosilicate glasses. *Contributions to Mineralogy and Petrology* **91**, 105–21.

Fine, G. & E. Stolper 1986. Dissolved carbon dioxide in basaltic glasses: Concentrations and speciation. *Earth and Planetary Science Letters* **76**, 263–278.

Finnerty, T. A., G. A. Waychunas, & W. M. Thomas 1978. The preparation of starting mixes for mineral synthesis by a freeze-dry technique. *American Mineralogist* **63**, 415–8.

Flynn, R. T. & C. W. Burnham 1978. An experimental determination of rare earth partition coefficients between a chloride containing vapor phase and silicate melts. *Geochimica Cosmochimica Acta* **42**, 685–701.

Ford, C. E. 1972. Furnace design, temperature distribution, calibration and seal design in internally heated pressure vessels. *Progress in Experimental Petrology. Natural Environmental Research Council (Great Britain) Series D*, **2**(11), 89–96.

Ford, C. E., D. G. Russell, J. A. Craven, & M. R. Fisk 1983. Olivine–liquid equilibrium. Temperature, pressure and composition dependence of the crystal–liquid cation partition coefficients for Mg, Fe^{2+}, Ca and Mn. *Journal of Petrology* **24**, 256–65.

Fournier, R. O. & J. J. Rowe 1966. Estimation of underground temperatures from the silica content of water from hot springs and wet-steam wells. *American Journal of Science* **264**, 685–97.

Franck, E. U. (1956). Hochverdichteter Wasserdampf I. Elektrolytische Leitfahigkeit in $KCl–H_2O$ Losungen bis $750°C$. *Zeitschrift für Physikalische Chemie* **8**, 92–106.

Franck, E. U. 1973. Concentrated electrolyte solutions at high temperature and pressure. *Journal of Solution Chemistry* **2**, 339–53.

Franck, E. U. 1974. Polar and ionic fluids at high pressures and temperatures. *Pure and Applied Chemistry* **38**, 449–68.

Franck, E. U., M. Brose, & K. Mangold 1962. Super critical hydrogen chloride. Specific heat up to $300°C$ and 300 atm. *PVT*-data up to $400°C$ and 2000 atm. *Progress in International Research on Thermodynamic Transport Properties. Papers from the Symposium on Thermophysical Properties*, 2nd edn, 159–65. Princeton, N.J.: Princeton University Press.

Frantz, J. D. & H. P. Eugster 1973. Acid–base buffers: Use of Ag + AgCl in the control of solution equilibria at elevated pressures and temperatures. *American Journal of Science* **267**, 268–86.

Frantz, J. D. & W. L. Marshall 1982. Electrical conductances and ionization constants of calcium chloride and magnesium chloride in aqueous solutions at temperatures to $600°C$ and pressures to 4000 bars. *American Journal of Science* **282**, 1666–93.

Frantz, J. D. & R. K. Popp 1979. Mineral solution equilibria. I. An experimental study of complexing and thermodynamic properties of aqueous MgCl in the system $MgO–SiO_2–H_2O–HCl$. *Geochimica Cosmochimica Acta* **43**, 1233–39.

Frantz, J. D., R. K. Popp, & N. Z. Boctor 1981. Mineral–solution equilibria – V. Solubilities of rock-forming minerals in supercritical fluids. *Geochimica Cosmochimica Acta* **45**, 69–77.

Fujii, T. & C. M. Scarfe 1985. Composition of liquids coexisting with spinel lherzolite at 10 kbar and the genesis of MORBS. *Contributions to Mineralogy and Petrology* **90**, 18–28.

Fyfe, W. S. 1960. Hydrothermal synthesis and determination of equilibrium between minerals in the subliquidus region. *Journal of Geology* **68**, 553–66.

Gasparik, T. 1984a. Experimental study of subsolidus phase relations and mixing properties of pyroxene in the system $CaO-Al_2O_3-SiO_2$. *Geochimica Cosmochimica Acta* **48**, 2537–45.

Gasparik, T. 1984b. Experimentally determined stability of clinopyroxenes + garnet + corundum in the system $CaO-MgO-Al_2O_3-SiO_2$. *American Mineralogist* **69**, 1025–35.

Gasparik, T. & R. C. Newton 1984. The reversed alumina contents of orthopyroxene in equilibrium with spinel and forsterite in the system $MgO-Al_2O_3-SiO_2$. *Contributions to Mineralogy and Petrology* **85**, 186–96.

Ghiorso, M. S. & I. S. E. Carmichael 1980. A regular solution model for metaluminous silicate liquids: Applications to geothermometry, immiscibility, and the source regions of basic magmas. *Contributions to Mineralogy and Petrology* **71**, 323–42.

Giggenbach, W. F. 1971. A simple spectrophotometric cell for use with aqueous solutions up to $280°C$. *Journal of Physics Series E, Scientific Instruments* **4**, 148–9.

Goldschmidt, V. M. 1911. Die Kontaktmetamorphose im Kristianiagebiet. *Kristina Vidensk Skritten* **1**, *Mathematics – Nature* **2**, 1–83.

Goldsmith, J. R. 1980. The melting and breakdown of reactions of anorthite at high pressures and temperatures. *American Mineralogist* **65**, 272–84.

Goldsmith, J. R. & D. M. Jenkins 1985. The high–low albite relations revealed by reversal of degree of order at high pressures. *American Mineralogist* **70**, 911–23.

Green, D. H. 1976. Experimental testing of "equilibrium" partial melting of peridotite under water-saturated, high pressure conditions. *Canadian Mineralogist* **14**, 255–68.

Green, D. H. & A. E. Ringwood 1967. The genesis of basaltic magmas. *Contributions to Mineralogy and Petrology* **15**, 103–90.

Grove, T. L. & W. B. Bryan 1983. Fractionation of pyroxene–phyric MORB at low pressure: An experimental study. *Contributions to Mineralogy and Petrology* **84**, 293–309.

Gunter, W. D., J. Myers, & J. R. Wood 1979. The Shaw bomb, an ideal hydrogen sensor. *Contributions to Mineralogy and Petrology* **70**, 23–7.

Gunter, W. D., J. D. Myers, & S. Girsperger 1987. Hydrogen: metal membranes. In *Hydrothermal experimental techniques*, Ulmer, G. C. & H. L. Barnes Jr. (eds.), 100–20. New York: Wiley.

Haas, H. & M. J. Holdaway 1973. Equilibria in the system $Al_2O_3-SiO_2-H_2O$ involving the stability limits of pyrophyllite, and thermodynamic data of pyrophyllite. *American Journal of Science* **273**, 449–64.

Hall, H. T. 1960. High pressure apparatus. In *Progress in very high pressure research*, F. P. Bundy, W. R. Hubbard Jr, & H. M. Strong (eds.), 1–10. New York: Wiley.

Hamilton, D. L. & C. M. B. Henderson 1968. The preparation of silicate compositions by a gelling method. *Mineralogical Magazine* **36**, 832–8.

Hamilton, D. L., C. W. Burnham & E. F. Osborn 1964. The solubility of water and effects of oxygen fugacity and water content on crystallization in mafic magmas. *Journal of Petrology* **5**, 21–39.

Harrison, W. J. & B. J. Wood 1980. An experimental investigation of the partitioning of REE between garnet and liquid with reference to the role of defect equilibria. *Contributions to Mineralogical Petrology* **72**, 145–55.

Hart, S. R. 1981. Diffusion compensation in natural silicates. *Geochimica Cosmochimica Acta* **45**, 279–92.

Hart, S. R. & K. E. Davis 1978. Nickel partitioning between olivine and silicate melt. *Earth and Planetary Science Letters* **40**, 203–19.

Heinz, D. L. & R. Jeanloz 1984. Compression of the B_2 high pressure phase of NaCl. *Physical Review Series B* **30**, 6045–50.

Helgeson, H. C., J. M. Delany, H. W. Nesbitt & D. K. Bird 1978. Summary and critique of the thermodynamic properties of rock-forming minerals. *American Journal of Science* **278**-A, 1–229.

Hemley, J. J., J. W. Montoya, D. R. Shaw, & R. W. Luce 1977. Mineral equilibria in the

BIBLIOGRAPHY

$MgO-SiO_2-H_2O$ system: II Talc–antigorite–forsterite–anthophyllite–enstatite stability relations and some geologic implications. *American Journal of Science* **277**, 353–83.

Henry, D. J., A. Navrotsky, & H. D. Zimmermann 1982. Thermodynamics of plagioclase melt equilibria in the system albite–anorthite–diopside. *Geochimica Cosmochimica Acta* **46**, 381–91.

Hettiarachi, S. & D. D. Macdonald 1983. Ceramic membranes for precise pH measurements in high temperature aqueous environments. *Journal of the Electrochemical Society* **131**, 2206–7.

Holland, T. J. B. 1980. The reaction albite $\rightleftharpoons$ jadeite + quartz determined experimentally in the range of $600-1200°C$. *American Mineralogist* **65**, 129–34.

Holland, T. J. B. 1983. The experimental determination of activities in disordered and short range ordered jadeitic pyroxenes. *Contributions to Mineralogy and Petrology* **82**, 214–20.

Holloway, J. R. 1971. Internally heated pressure vessels. In *Research techniques for high temperature and pressure*, G. C. Ulmer (ed.), 217–257. New York: Springer.

Holloway, J. R. 1973. The system paragasite–H_2O–CO_2: A model for melting of a hydrous mineral with a mixed-volatile fluid – I. Experimental results to 8 kbars. *Geochimica Cosmochimica Acta* **37**, 651–66.

Holloway, J. R. 1987. Igneous fluids. In *Thermodynamic modelling of geological materials: minerals, fluids and melts*, I. S. E. Carmichael & H. P. Eugster (eds.). *Reviews in Mineralogy*, **17**, 211–34.

Holloway, J. R., C. W. Burnham & G. L. Millhollen 1968. Generation of H_2O–CO_2 mixtures for use in hydrothermal experimentation. *Journal of Geophysical Research* **73**, 6598–600.

Holloway, J. R. & R. L. Reese 1974. The generation of N_2–CO_2–H_2O fluids for use in hydrothermal experimentation. I. Experimental methods and equilibrium calculation in the C–O–H–N system. *American Mineralogist* **59**, 589–97.

Huebner, J. S. 1971. Buffering techniques for hydrostatic systems at elevated pressures. In *Research techniques for high pressure and high temperature*, G. C. Ulmer (ed.), 123–77. New York: Springer.

Huebner, J. S. 1987. Use of gas mixtures at low pressure to specify oxygen and other fugacities of furnace atmospheres. In *Hydrothermal experimental techniques*, G. C. Ulmer & H. L. Barnes Jr. (eds.), 20–60. New York: Wiley.

Hultgren, R., P. D. Desa, D. T. Hawkins, M. Gleiser, & K. K. Kelley 1973. Selected values of the thermodynamic properties of binary alloys. Metals Park, Ohio: America Society for Metals.

Hultgren, R., P. Newcomb, R. L. Orr, & L. Warner 1959. A diphenyl ether calorimeter for measuring high temperature heat contents of metals and alloys. Proceedings of the 9th National Physical Laboratories Symposium. London: Her Majesty's Stationery Office.

Irving, A. J. & P. J. Wyllie 1975. Subsolidus and melting relationships for calcite, magnesite on the join $CaCO_3$–$MgCO_3$ to 36 kb. *Geochimica Cosmochimica Acta* **39**, 35–53.

Ito, E., E. Takahashi, & Y. Matsui 1984. The mineralogy and chemistry of the lower mantle: An implication of the ultrahigh-pressure phase relations in the system MgO–FeO–SiO_2. *Earth and Planetary Sciences Letters* **67**, 238–48.

Jakobsson, S. & J. R. Holloway 1986. Crystal–liquid experiments in the presence of a C–O–H fluid buffered by graphite + iron + wustite: Experimental method and near-liquidus relations in basanite. *Journal of Volcanology and Geothermal Research* **29**, 265–91.

Jeanloz, R. & A. B. Thompson 1983. Phase transitions and mantle discontinuities. *Reviews of Geophysics and Space Physics* **4**, 51–74.

Jenkins, D. M., J. R. Holloway, & J. F. Kacoyannakis 1984. Temporal variation of aqueous constituents in a water–basalt–supercalcine system: Implications for the experimental assessment of nuclear waste forms. *Geochimica Cosmochimica Acta* **48**, 1443–54.

Jenkins, D. M. & R. C. Newton 1979. Experimental determination of the spinel peridotite to

187

garnet peridotite inversion at $900°C$ and $1000°C$ in the system $CaO-MgO-Al_2O_3-SiO_2$, and at $900°C$ with natural garnet and olivine. *Contributions to Mineralogy and Petrology* **68**, 407–19.

Jephcoat, A. P., H.-K. Mao, & P. M. Bell 1987. Operation of the megabar diamond anvil cell. In *Hydrothermal experimental techniques*, G. C. Ulmer & H. L. Barnes Jr. (eds.), 469–506. New York: Wiley.

Johannes, W. 1978. Melting of plagioclase in the systems $Ab-An-H_2O$, $Qz-Ab-An-H_2O$ at $P_{H_2O} = 5$ bar. An equilibrium problem. *Contributions to Mineralogy and Petrology* **66**, 295–303.

Johannes, W., P. M. Bell, H. K. Mao, A. L. Boettcher, D. W. Chipman, J. F. Hays, R. C. Newton, & F. Selfert 1971. An interlaboratory comparison of piston-cylinder pressure calibration using the albite breakdown reaction. *Contributions to Mineralogy and Petrology* **32**, 24–38.

Johannes, W. & D. Puhan 1971. The calcite–aragonite transition, reinvestigated. *Contributions to Mineralogy and Petrology* **31**, 28–38.

Kalinina, A. M., V. N. Filipovich, & V. M. Fokin 1980. Stationary and non-stationary crystal nucleation rate in a glass of $2Na_2O \cdot CaO \cdot 3SiO_2$ stoichiometric composition. *Journal of non-Crystalline Solids 38/39*, 723–8.

Karsten, J. L., J. R. Holloway, & J. R. Delaney 1982. Ion microprobe studies of water in silicate melts: Temperature-dependent water diffusion in obsidian. *Earth and Planetary Sciences Letters* **59**, 420–28.

Kerrick, D. M. 1987. Cold-seal systems. In *Hydrothermal experimental techniques*, G. C. Ulmer & H. L. Barnes Jr. (eds.), 293–323. New York: Wiley.

Kilinc, A., I. S. E. Carmichael, M. L. Rivers, & R. O. Sack 1983. The ferric–ferrous ratio of natural silicate liquids equilibrated in air. *Contributions to Mineralogy and Petrology* **83**, 136–40.

Kilinc, I. A. & C. W. Burnham 1972. Partitioning of chloride between a silicate melt and coexisting aqueous phase from 2 to 8 kilobars. *Economic Geology* **67**, 231–35.

Kirkpatrick, R. J. 1981. Kinetics of crystallization of igneous rocks. In *Reviews in Mineralogy*, A. C. Lasaga & R. J. Kirkpatrick (eds.), **8**, 321–98. Washington D.C.: Mineralogical Society of America.

Kleppa, O. J. 1960. A new twin type high temperature reaction calorimeter. The heat of mixing in liquid sodium–potassium metals. *Journal of Physics and Chemistry* **64**, 1937–40.

Kleppa, O. J. 1972. Oxide melt solution calorimetry. *Colloque International du CNRS, Thermochimie,* **201**, 119–27.

Kleppa, O. J. 1976. Mineralogical applications of high temperature reaction calorimetry. In *The physics and chemistry of minerals and rocks*, R. G. Strens (ed.), London: Wiley.

Krupka, K. M., R. A. Robie & B. S. Hemingway 1979. High temperature heat capacities of corundum, periclase, anorthite, $CaAl_2Si_2O_8$ glass, muscovite, anthophyllite, $KAlSi_3O_8$ glass, grossular and $NaAlSi_3O_8$ glass. *American Mineralogist* **64**, 86–101.

Kuikkola, K. & C. Wagner 1957. Measurement of galvanic cells including solid electrolytes. *Journal of the Electrochemical Society* **104**, 379–87.

Kushiro, I. 1976. Changes in viscosity and structure of melt of $NaAlSi_2O_6$ composition at high pressures. *Journal of Geophysical Research* **81**, 6347–50.

Lapham, K. E., J. R. Holloway & J. D. Delaney 1984. Diffusion of H_2O and D_2O in obsidian at elevated temperatures and pressures. *Journal of Non-Crystalline Solids* **67**, 179–91.

Lasaga, A. C. 1981. Rate laws of chemical reactions. *Reviews in Mineralogy* **8**, 321–97.

Lasaga, A. C. 1984. Chemical kinetics of water–rock interactions. *Journal of Geophysical Research* **89**, 4009–25.

Liebermann, R. C., C. T. Prewitt, & D. J. Weidner 1985. Large-volume high-pressure mineral physics in Japan. *Eos, Transactions, American Geophysical Union* **66**, 138–39.

Lofgren, G. 1987. Internally heated systems. In *Hydrothermal Experimental Techniques*, G. C. Ulmer & H. L. Barnes, Jr. (eds.), 324–32. New York: Wiley.

Luth, W. C. & O. F. Tuttle 1963. Externally heated cold-seal pressure vessels for use to 10,000 bars and 750°C. *American Mineralogist* **48**, 1401–3.

Macdonald, D. D., A. C. Scott, & P. Wentrock 1981. Redox potential measured in high temperature aqueous systems. *Journal of the Electrochemical Society* **128**, 250–7.

Macdonald, D. D., P. Wentrock, & A. C. Scott 1980. The measurement of pH in aqueous systems at elevated temperature using palladium hydride electrodes. *Journal of the Electrochemical Society* **127**, 1745–51.

Margaritz, M. & A. W. Hoffman 1978. Diffusion of Sr, Ba and Na in obsidian. *Geochimica Cosmochimica Acta* **42**, 595–605.

McMillan, P. 1984. Structural studies of silicate glasses and melts – Applications and limitation of raman spectroscopy. *American Mineralogist* **69**, 622–44.

McMillan, P. F. & J. R. Holloway 1987. Water solubility in aluminosilicate melts. *Contributions to Mineralogy and Petrology* **97**, 320–32.

Mirwald, P. W., I. C. Getting, & G. C. Kennedy 1975. Low-friction cell for piston-cylinder high-pressure apparatus. *Journal of Geophysical Research* **80**, 1519–25.

Mukherjee, S. P. 1984. Inorganic oxide gels and gel monoliths: Their crystallization behavior. *Material Science Research* **17**, 95–109.

Mysen, B. O. & M. Seitz 1975. Experimental determination of distributions and concentrations of trace elements up to the physical conditions corresponding to the upper mantle – C and Sm as examples. *Journal of Geophysical Research* **80**, 2627–35.

Nafziger, R. H., G. C. Ulmer, & E. Woermann 1971. Gaseous buffering for the control of oxygen fugacity at one atmosphere. In *Techniques for high pressure and high temperature*, G. C. Ulmer (ed.), 9–41. New York: Springer.

Nakamura, A. & H. Schmalzried, 1983. On the nonstoichiometry and point defects of olivine. *Physics and Chemistry of Minerals* **10**, 27–37.

Nekvasil, H. & C. W. Burnham 1987. The calculated individual effects of pressure and water content on phase equilibria in the granite system. In *Magmatic processes: physicochemical principles*, B. O. Mysen (ed.), The Geochemical Society Special Publication **1**, 433–45.

Newton, R. C. 1966. Kyanite–sillimanite equilibrium at 750°C. *Science* **151**, 1222–25.

Newton, R. C. & B. J. Wood 1980. Volume behavior of silicate solid solutions. *American Mineralogist* **65**, 733–45.

Newton, R. C., T. V. Charlu, & O. J. Kleppa 1980. Thermochemistry of the high structural state plagioclases. *Geochimica Cosmochimica Acta* **44**, 933–41.

Niedrach L. 1980. A new membrane-type pH sensor for use in high temperature–high pressure water. *Journal of the Electrochemical Society* **127**, 2122–30.

Nielsen, R. L. & M. A. Dungan 1983. Low pressure mineral–melt equilibria in natural anhydrous mafic systems. *Contributions to Mineralogy and Petrology* **84**, 310–26.

O'Neill, H. St. C. & B. J. Wood 1979. An experimental study of Fe–Mg partitioning between garnet and olivine and its calibration as a geothermometer. *Contributions to Mineralogy and Petrology* **70**, 59–70.

Osborn, E. F. 1959. Role of oxygen pressure in the crystallization and differentiation of basaltic magma. *American Journal of Science* **47**, 211–26.

Poty, B., H..A. Stadler, & A. M. Weisbrod 1974. Fluid inclusion studies in quartz from fissures of western and central Alps. *Schweizerische Mineralogische und Petrografische Mitteilung* **54**, 717–52.

189

Presnall, D. C., J. R. Dixon, T. H. O'Donnell, N. L. Brenner, R. J. Schrock & D. W. Dycus 1978. Liquidus phase relations on the join diopside–forsterite–anorthite from 1 atm. to 29 kbar.: Their bearing on the generation and crystallization of basaltic magma. *Contributions to Mineralogy and Petrology* **66**, 203–20.

Quist, A. S. & W. L. Marshall 1968. Electrical conductances of aqueous sodium chloride from $0°$ to $800°$ and at pressures to 4000 bars. *Journal of Physical Chemistry* **72**, 684–703.

Ragnarsdottir, K. V. & J. V. Walther 1983. Pressure sensitive "silica geothermometer" determined from quartz solubility experiments at $250°C$. *Geochimica Cosmochimica Acta* **47**, 941–46.

Read, H. H. 1948. Granites and granites. In *Origin of granite: a symposium*. Geological Society of America Memorandum **28**, 1–19.

Richet, P. & Y. Bottinga 1984. Anorthite, andesine, wollastonite, diopside, cordierite, and pyrope: thermodynamics of melting, glass transitions, and properties of the amorphous phases. *Earth and Planetary Sciences Letters* **67**, 415–32.

Rigden, S. M., T. J. Ahrens, & E. M. Stolper 1984. Densities of liquid silicates at high pressures. *Science* **226**, 1071–74.

Rimstidt, J. D. & H. L. Barnes 1980. The kinetics of silica–water reactions. *Geochimica Cosmochimica Acta* **44**, 1683–99.

Robie, R. A. 1987. Calorimetry. In *Hydrothermal experimental techniques*, G. C. Ulmer & H. L. Barnes Jr. (eds.), 389–422. New York: Wiley.

Robie, R. A. & B. S. Hemingway 1972. *Calorimeters for heat of solution and low-temperature heat capacity measurements*. U.S. Geological Survey Professional Paper, **755**.

Robie, R. A., B. S. Hemingway & W. H. Wilson 1978. Low-temperature heat capacities and entropies of feldspar glasses and of anorthite. *American Mineralogist* **63**, 109–23.

Roeder, P. L. & R. F. Emslie 1970. Olivine–liquid equilibrium. *Contributions to Mineralogy and Petrology* **29**, 275–89.

Rudert, V., I. Chou, & H. P. Eugster 1976. Temperature gradients in rapid-quench cold-seal pressure vessels. *American Mineralogist* **61**, 1012–15.

Sato, M. 1971. Electrochemical measurements and control of oxygen fugacity and other gaseous fugacities with solid electrolyte sensors. In *Research techniques for high pressure and high temperature*, G. C. Ulmer (ed.), 43–99. New York: Springer.

Schairer, J. F. 1951. Phase transformations in polycomponent silicate systems. In *Phase transformations in solids*, J. E. Smoluchowski, J. E. Mayer, & W. A. Weyl (eds.), 278–95. New York: Wiley.

Schairer, J. F. 1959. Phase equilibria with particular reference to silicate systems. In *Physicochemical measurements at high temperatures*, 117–34. London: Butterworth.

Seward, T. M. 1976. The stability of chloride complexes in hydrothermal solutions up to $350°C$ *Geochimica Cosmochimica Acta* **40**, 1329–41.

Seward, T. M. 1981. Metal complex formation in aqueous solutions. In *Physics and Chemistry of the Earth 13*, D. T. Richard & F. E. Wickman (eds.), New York: Pergamon.

Seward, T. M. 1984. The formation of lead (II) chloride complexes to $300°C$: A Spectrophotometric study. *Geochimica Cosmochimica Acta* **48**, 121–34.

Seward, T. M. & E. U. Franck 1981. The system hydrogen–water up to $440°C$ and 2500 bar pressure. *Berichte der Bunsen-Gesellschaft für Physikalische Chemie* **86**, 2–7.

Seyfried, W. E. & J. L. Bischoff 1979. Low temperature basalt interaction by seawater: An experimental study at $700°C$ and $150°C$. *Geochimica Cosmochimica Acta* **43**, 1937–47.

Seyfried, W. E. & D. R. Janecky 1985. Heavy metal and sulfur transport during subcritical and supercritical hydrothermal alteration of basalt: Influence of fluid pressure and basalt composition and crystallinity. *Geochimica Cosmochimica Acta* **49**, 2545–60.

BIBLIOGRAPHY

Seyfried, W. E., Jr., P. C. Gordon, & F. W. Dickson 1979. A new reaction cell for hydrothermal solution equipment. *American Mineralogist* **64**, 646–9.

Seyfried, W. E., Jr., D. R. Janecky, & M. E. Berndt 1987. Rocking autoclaves for hydrothermal experiments II. The flexible reaction cell system. In *Hydrothermal experimental techniques*, G. C. Ulmer & H. L. Barnes, Jr. (eds.), 216–39. New York: Wiley.

Shaw, H. R. 1963a. Hydrogen–water vapor mixtures: Control of hydrothermal atmospheres by hydrogen osmosis. *Science* **139**, 1220–22.

Shaw, H. R. 1963b. Obsidian–H_2O viscosities at 1000 and 2000 bars in the temperature range $700°C$ to $900°C$. *Journal of Geophysical Research* **68**, 6337–43.

Shaw, H. R. 1967. Hydrogen osmosis in hydrothermal experiments. In *Researches in Geochemistry*, P. H. Abelson (ed.), **2**, 521–41. New York: Wiley.

Shettel, D. L. 1974. The solubility of quartz in supercritical H_2O–CO_2 fluids. M.S. Thesis, Pennsylvania State University.

Skippen, G. B. 1971. Experimental data for reactions in siliceous marbles. *Journal of Geology* **79**, 457–81.

Skippen, G. B. 1974. An experimental model for low pressure metamorphism of siliceous dolomitic marble. *American Journal of Science* **274**, 487–509.

Slaughter, J., D. M. Kerrick, & V. J. Wall 1975. Experimental and thermodynamic study of equilibria in the system CaO–MgO–SiO_2–H_2O–CO_2. *American Journal of Science* **275**, 143–62.

Spetzler, H. 1970. Equation of state of polycrystalline and single-crystal Mgo to 8 kilobars and $800°K$. *Journal of Geophysical Research* **75**, 2073–87.

Stolper, E. M. 1982. Water in silicate glasses: an infrared spectroscopic study. *Contributions to Mineralogy and Petrology* **81**, 1–17.

Stolper, E. M., D. Walker, & J. F. Hays 1980. Melt segregation from partially molten source regions: the importance of melt density and source region size. *Journal of Geophysical Research* **86**, 6161–71.

Takahashi, E. 1985. Melting of dry peridotite KLB 1 up to 14 Gpa: implications on the origin of peridotitic upper mantle. *Journal of Geophysical Research* **91**, B9367–82.

Takahashi, E. & I. Kushiro 1983. Melting of a dry peridotite at high pressures and basalt magma genesis. *American Mineralogist* **68**, 859–79.

Thornton, E. C. & W. E. Seyfried 1985. Sediment–sea-water interaction at 200 and $300°C$, 500 bars pressure: Low grade metamorphism of marble clay. *Geological Society of America Bulletin* **96**, 1287–95.

Tilley, C. E. 1925. Metamorphic zones in the Southern Highlands of Scotland. *Geological Society of London Quarterly Journal* **81**, 100–12.

Torgeson, D. R. & T. G. Sahama 1948. Hydrofluoric acid solution calorimeter and the determination of the heats of formation of Mg_2SiO_4, $MgSiO_3$, and $CaSiO_3$. *Journal of the American Chemical Society* **70**, 2156–60.

Tuttle, O. F. 1949. Two pressure vessels for silicate–water studies. *Geological Society of America Bulletin* **60**, 1727–9.

Tuttle, O. F. & N. L. Bowen 1958. Origin of granite in the light of experimental studies in the system $NaAlSi_3O_8$–$KAlSi_3O_8$–SiO_2–H_2O. *Geological Society of America Memoir* **74**.

Tyburczy, J. A., B. Frisch, & T. J. Ahrens 1986. Shock-induced volatile loss from a carbonaceous chondrite: Implications for planetary accretion. *Earth and Planetary Sciences Letters* **80**, 201–7.

Urbain, G., Y. Bottinga, & P. Richet 1982. Viscosity of liquid silica, silicates and aluminosilicates. *Geochimica Cosmochimica Acta* **46**, 1061–72.

van Valkenburg, A. 1963. Visual observations of high-pressure transitions. *Reviews of Scientific Instruments* **33**, 1462.

191

van Valkenburg, A., P. M. Bell & H.-K Mao 1987. High-pressure mineral solubility experiments in the diamond-window cell. In *Hydrothermal experimental techniques*, G. C. Ulmer & H. L. Barnes Jr. (eds.), 458–468. New York: Wiley.

Walker, D., T. Shibata, & S. E. Delong 1979. Abyssal tholeiites from the Oceanographer fracture zone. *Contributions to Mineralogy and Petrology* **70**, 111–25.

Walsh, J. M. & R. H. Christian 1955. Equation of state of metals from shock-wave measurements. *Physical Reviews* **97**, 1544–56.

Walther, J. V. & P. M. Orville 1983. The extraction–quench technique for determination of the thermodynamic properties of solute complexes: Application to quartz solubility in fluid mixtures. *American Mineralogist* **68**, 731–41.

Watson, E. B. 1977. Partitioning of manganese between forsterite and silicate liquid. *Geochimica Cosmochimica Acta* **41**, 1363–74.

Watson, E. B. 1979a. Calcium diffusion in a simple silicate melt to 30 kbar. *Geochimica Cosmochimica Acta* **43**, 313–22.

Watson, E. B. 1979b. Zircon saturation in felsic liquids: Experimental results and applications to trace element geochemistry. *Contributions to Mineralogy and Petrology* **70**, 407–19.

Watson, E. S., M. J. O'Neill, J. Justin, & W. N. Brent 1964. A differential scanning calorimeter for quantitative differential thermal analysis. *Analytical Chemistry* **36**, 1233–8.

Weidner, D. J., K. Swyler, & H. R. Carleton 1975. Elasticity of microcrystals. *Geophysical Research Letters* **2**, 189–92.

Weill, D. F. 1966. Stability relations in the Al_2O_3–SiO_2 system calculated from solubilities in the Al_2O_3–SiO_2–Na_3AlF_6 system. *Geochimica Cosmochimica Acta* **30**, 223–37.

Weill, D. F., R. Hon & A. Navrotsky 1980. The igneous system $CaMgSi_2O_6$–$CaAl_2Si_2O_8$ –$NaAlSi_3O_8$: variations on a classic theme by Bowen. In *Physics of Magmatic Processes*, R. B. Hargraves (ed.), 49–92.

Weir, C. E., E. R. Lippincott, A. Van Valkenburg & E. N. Bunting 1959. Infrared studies in the 1- to 15-micron region to 30,000 atmospheres. *Journal of Research of the National Bureau of Standards, Section A,* **63**, 55–62.

Westrum, E. F., Jr., G. T. Furukawa & J. P. McCullough 1968a. Adiabatic low temperature calorimetry. In *Experimental thermodynamics vol. I. Calorimetry of non-reacting systems*, J. P. McCullough & D. W. Scott (eds.), 133–214. London: Butterworth.

Westrum, E. F., Jr., R. R. Walters, H. E. Flotow, & D. W. Osborne 1968b. Uranium monosulfide-ferromagnetic transition-heat capacity and thermodynamic properties from 1.5 to $350°K$ *Journal of Chemistry and Physics* **48**, 155–61.

Whitney, J. A. & J. C. Stormer 1976. Geothermometry and geobarometry in epizonal granitic intrusions: A comparison of iron–titanium oxides and coexisting feldspars. *American Mineralogist* **61**, 751–61.

Williams, D. W. 1968. Improved cold seal pressure vessels to operate to $1100°C$ at 3 kilobars. *American Mineralogist* **53**, 1765–9.

Winchell, P. 1969. The compensation law for diffusion in silicates. *High Temperature Science* **1**, 200–15.

Wood, B. J. 1977. Experimental determination of the mixing properties of solid solutions with particular reference to garnet and clinopyroxene solutions. In *Thermodynamics in geology*, D. G. Fraser (ed.), 11–27. Dordrecht, Holland: D. Reidel.

Wood, B. J. 1987. Thermodynamics of multicomponent systems containing several solid solutions. *Reviews in Mineralogy* **17**, 71–94.

Wood, B. J. & D. G. Fraser 1976. *Elementary thermodynamics for geologists*. Oxford: Oxford University Press.

Wood, B. J., T. J. B. Holland, R. C. Newton & O. J. Kleppa 1980. Thermochemistry of jadeite–diopside pyroxenes. *Geochimica Cosmochimica Acta* **44**, 1363–71.

Wood, B. J. & J. R. Holloway 1982. Theoretical prediction of phase relationships in planetary mantles. *Journal of Geophysical Research* **87**, A19–A30.

BIBLIOGRAPHY

Wood, B. J. & J. R. Holloway 1984. A thermodynamic model for subsolidus equilibria in the system CaO–MgO–Al$_2$O$_3$–SiO$_2$. *Geochimica Cosmochimica Acta* **48**, 159–76.

Wood, B. J. & O. J. Kleppa 1981. Thermochemistry of forsterite–fayalite olivine solutions. *Geochimica Cosmochimica Acta* **45**, 529–34.

Wood, B. J. & J. V. Walther 1983. Rates of hydrothermal reactions. *Science* **222**, 413–5.

Wyllie, P. J. 1963. The quenching technique in non-quenchable systems: a discussion concerning the alleged thermal decomposition of portlandite at high pressures. *American Journal of Science* **261**, 983–8.

Xu, J., H.-K. Mao, & P. M. Bell 1986. High-pressure ruby and diamond fluorescence: Observations at 0.20–0.55 terapascals (2–5.5 megabars). *Science* **232**, 1401–6.

Yagi, T. & S. Akimoto 1982. X-ray measurements to 100 GPa range and static compression of α–Fe$_2$O$_3$. In *High pressure research in geophysics*, S. Akimoto & M. H. Manghnami (eds.), Dordrecht, Holland: D. Reidel.

Yagi, T. & S. Akimoto 1984. *In situ* observations of crystal structure and phase transformations under pressure. In *Materials science of the Earth's interior*, I. Sunagawa (ed.), Dordrecht, Holland: D. Reidel.

Yoder, H. S., Jr. 1950. Stability relations of grossularite. *Journal of Geology* **58**, 221–53.

Yoder, H. S., Jr. & C. E. Tilley 1962. Origin of basalt magmas: An experimental study of natural and synthetic systems. *Journal of Petrology* **3**, 342–532.

Zimmerman, H. D., J. R. Holloway & A. Navrotsky 1985. Plagioclase–melt equilibria in the system albite–anorthite–diopside: New direct measurements and thermodynamic calculations. *Eos, Transactions, American Geophysical Union* **66**, 412.

Index